# The h[illegible] tech entrepreneur's handbook

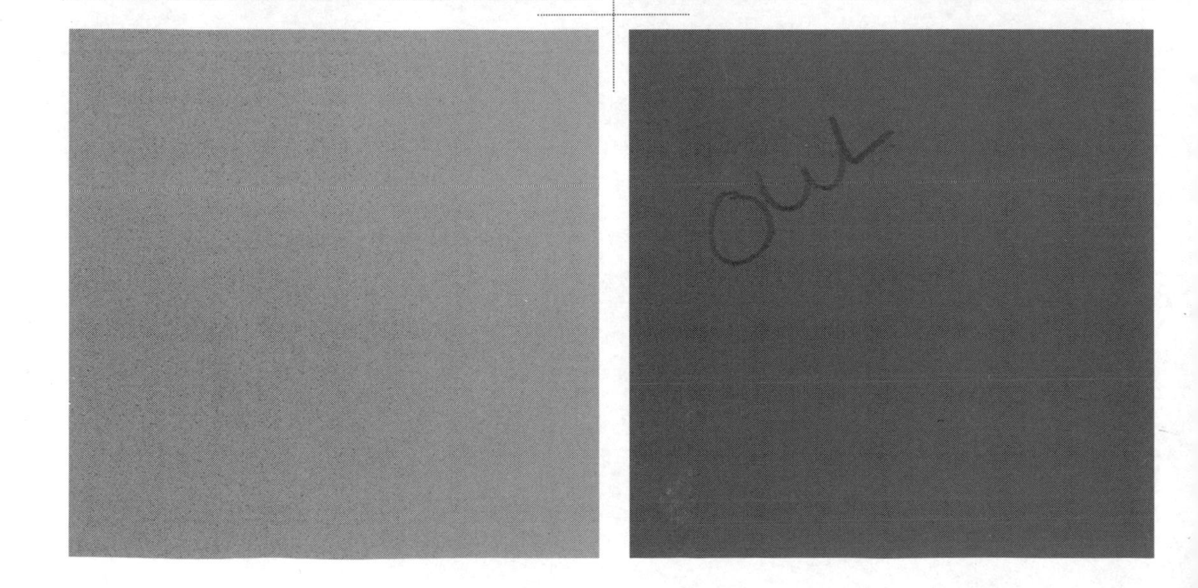

# The high-tech entrepreneur's handbook

HOW TO START AND RUN A HIGH-TECH COMPANY

JACK LANG
AND THE CAMBRIDGE ENTREPRENEURSHIP CENTRE

London · New York · Toronto · Sydney · Tokyo · Singapore · Hong Kong · Cape Town
New Delhi · Madrid · Paris · Amsterdam · Munich · Milan · Stockholm

**PEARSON EDUCATION LIMITED**

Edinburgh Gate
Harlow CM20 2JE
Tel: +44 (0)1279 623623
Fax: +44 (0)1279 431059
Website: www.pearsoned.co.uk

First published in Great Britain in 2002



ISBN: 0 273 65615 5

*British Library Cataloguing in Publication Data*
A CIP catalogue record for this book can be obtained from the British Library.



10 9 8 7 6 5 4 3

Designed by Design Deluxe, Bath
Typeset by Pantek Arts Ltd, Maidstone, Kent
Printed and bound in Great Britain by Biddles Ltd,
*www.biddles.co.uk*

The Publishers' policy is to use paper manufactured from sustainable forests.

# Acknowledgements

THIS BOOK WOULD NOT HAVE BEEN POSSIBLE WITHOUT the help and support of many people, and in particular those at the Cambridge Entrepreneurship Centre. Peter Hiscocks, the Centre's Director, wrote most of the sections on idea generation and company valuation; Dr Tim Minshall the case studies and examples. Dr Joanna Harrison acted as a project co-ordinator. Sarah Snyder and Shirley Jamieson helped pull together the structure and content, and Liz Eaton assisted with the detailed editing. I also need to thank my publisher, Rachael Stock and her team at Pearson Education, for making difficult things seem easy. Other people read and commented on parts of the drafts, both within the Centre and externally, including Nicholas Bohm and Michael Carter (CFO extraordinary), making it a better book. The faults remain entirely mine.

# Example and disclaimer

IN CASE YOU HAVEN'T NOTICED, THE COMPANY USED as an example in Chapter 17, CEC Design Ltd, manufacturers of fine holographic projectors, is purely fictitious, and no resemblance is intended with any company, group or person, or with any technology now or in the future. Similarly, no reliance should be placed on examples of documents. They are intended to be illustrative rather than definitive.

# Support sites

Additional material can be found on:

**www.cec.cam.ac.uk**
**www.business-minds.com**

# Contents

## Marketing and selling 275

## Growth and exit 331

# Introduction

THIS BOOK HAS ARISEN FROM A HIGHLY SUCCESSFUL course of lectures in business studies given to students of the Computer Laboratory at the University of Cambridge. For the past 30 years or so, an average of roughly one graduate per year from the Laboratory has founded a company, and become a self-made multi-millionaire as a result.

One of the main themes of the original lecture series, and subsequently this book, is that technology, regardless of how good it is, forms only a small part of the process of bringing a project to a successful real-world conclusion, and plays an even smaller role in building a successful, reliable and useful enterprise. Although the book is based on, and mostly uses examples from, the world of high technology, the issues and lessons that it covers are universal, and apply to most enterprises.

‘ALTHOUGH THE BOOK IS BASED ON, AND MOSTLY USES EXAMPLES FROM THE WORLD OF HIGH TECHNOLOGY, THE ISSUES AND LESSONS THAT IT COVERS ARE UNIVERSAL, AND APPLY TO MOST ENTERPRISES’

The range of subject matter that this book covers is very large, and hence the coverage is broad rather than deep. It serves as an overview of the territory involved in starting and running a high-tech enterprise, and is intended to give an overall understanding, with pointers to further, more detailed sources of information. Thus you will not learn to be, for example, an accountant or patent lawyer from this book, but might gather enough to understand the principles involved, their importance, and how to work with, and understand the professional masters of the art.

The original course was for third-year undergraduates and post-graduates about to enter the world of work. The approach is essentially practical and hands-on. It is aimed at those starting out, for example those starting their own companies, or taking control of individual projects or departments within a larger company. It's about doing it day to day, rather than the more strategic issues that face higher management of global-scale corporations discussed in business schools, although the growth of internet commerce means that even smaller enterprises now have global reach.

# Getting started

# 1 So you've got an idea …

“**Dimidium facti qui coepit habet; sapere aude, incipe**. To have made a start is half of the business; dare to know, so begin!”

(Horace, **Epistles**, I:2) first century AD

## Introduction

OK, so you have this brilliant idea. You've spotted this real need or you've had an idea for this wonderful piece of technology, or your company has put you in charge of this great project. Now what are you going to do? How can you turn the idea into reality?

This chapter gives a framework and a context for the rest of the book. However, there is something of a chicken and egg problem. Some of the detail you will need is given in the rest of the book, so you might want to come back and re-read the early chapters when you have read through the book once.

## Why are you doing it?

This bit is about you. Before continuing, step back a moment and examine your own motives. Starting a company or running a significant project is going to take all your time and energy, and once the ball starts to roll you will have very little time for reflection. Now is a good time to figure out just why you want to do it, and what you want out of it, otherwise you are in danger of going to all that effort to end up with something you don't want.

It's also much easier to convince others if you have your own motivation sorted out. Setting some realistic personal goals will enable you to keep score on how well you are doing, and will provide a quick reality check. It's hard to discover your own goals, and it needs a lot of introspection but it is worthwhile. Be brutally honest with yourself. A clear set of personal objectives should also help avoid foolish choices, or at least make you consider some of the consequences.

Financial rewards are, of course, important, providing stability and motivation, and the lubrication of the machine. However, cash is not the only thing that motivates people, since lifestyle, other interests and ethical considerations are also important. You might validly decide that quality of life, or ethical considerations and getting a good night's sleep is more important to you than maximizing financial gain, in which case you should avoid high-stress, high-reward activities like day-trading or arms sales.

There are a number of things to consider.

## Fun or profit?

Do you want your company to be a lifestyle company, where the motivation is fun for the founders, who get to play with technological toys? Or do you want it to be designed primarily as a high-growth money-making machine, where the founders get their thrills from watching the share price go up? It's hard (but occasionally possible) to do both. Sometimes a lifestyle company will invent something for which there is great demand and strike gold, although that in itself is likely to change the nature of the company, and sometimes a well-managed money machine can be a great place to work.

‘THESE TWO TYPES OF COMPANY ARE VERY DIFFERENT BEASTS’

These two types of company are very different beasts, with different cultures. A lifestyle company will typically stay small, depend heavily on the founders, and grow organically using little external capital. Examples might be businesses such as small consultancies, specialist software companies, or even specialist retail shops. High-growth companies, by contrast, expand quickly, depend on bought-in management, and consume large amounts of external capital. The actual business area of the company may be only a professional interest, rather than the all-consuming passion of the founder.

## Size

Size matters. What size do you want your ideal company to be, now and in five years' time? Too small, and it may not be stable, as it will have to rely on a few key people, or a single contract or product. Small, and it may not be able to gain enough share of its niche market to keep out competitors. Yet, if it is large, it can become unwieldy and will need to suck in a lot of capital to fund the growth. You may have to give up significant shareholding and control to raise the money. Each time the company grows by roughly seven-fold (7, 50, 350, etc.) you will need to restructure and painfully introduce new layers of management, and so have less contact with people on the ground.

Of course, you might not have much control over the eventual size of the company, which will probably be more dependent on external factors such as success in the chosen market; it's hard to turn down opportunities as they arise. However when making key decisions, such as whether to introduce a new product, open a new shop or enter a new export market, remember your ideas about the size of company that you feel comfortable with.

## Timescale

Timescale is very important. Is this new activity the thing you want to do for the rest of your life? Is this the thing you want to be remembered for? Or is it just a quick way of earning some cash to fund a different lifestyle or to enable you pursue your real interests? If you are just motivated by short-term cash, and your real interests and life are elsewhere, are you doing the right thing? Maybe it would be better to become a trader, for example?

## Why me?

You need to be able to answer the question of what singles you out as being able to make your venture successful, compared to all the other hopefuls.

What unique advantages do you bring to the venture you propose? What's special about you? How can you capitalize on your strengths, and what do you need to do to compensate for your weaknesses?

Bear in mind that starting a business is a high-risk activity, and that most new ventures fail. It will take all your energy and waking hours, and some of the sleeping ones as well. It will be rough on your social relationships. Do you really want this? Are you prepared to gamble with your savings and your life? Do you have the support structures in place, like family and friends who will support you during the tough patches? Would you be better off and gain more power and influence, if that is what motivates you, by following a more conventional route, working your way up to a senior position in a large company, or pursuing a career in the city, politics, the civil service or academia?

Maybe you have a new idea, or unique knowledge about the market sector, or a proven track record of successful management. If you are ambitious, and moderately lucky, striking out on your own can be rewarding, both intellectually and financially. Be aware though, it is a hard and fraught road.

An inventor is not necessarily an entrepreneur. Entrepreneurs and inventors have different characters and skill sets. While an inventor may originate an idea, it takes different skills to build a team, raise the money and cope with all the nitty-gritty day–to-day details. Consider, for example, the differences between Alan Sugar, who started as a market trader, and Sir Clive Sinclair, the inventor. Both competed in the early home computer market, but the more commercially oriented Amstrad empire eventually took over Sinclair Research. It is possible for an inventor to become an entrepreneur, as James Dyson did, but as a result they often stop inventing. If your passion is invention, it might be better to sell the idea to an entrepreneur to develop, and use the cash to invent the next thing.

“IF YOUR PASSION IS INVENTION, IT MIGHT BE BETTER TO SELL THE IDEA TO AN ENTREPRENEUR TO DEVELOP, AND USE THE CASH TO INVENT THE NEXT THING”

You are unlikely to be able to do it all on your own. You will need to work with other people, with different skills, but also different ideals. They may be partners, or fellow directors, or employees. They may be in companies with whom you have forged an alliance. We discuss team building later in the book, but for now you might want to think about the sorts of skills and people you will need to recruit to complement your own talents. You might also want to think about how you are going to work together. It is much better to work out the terms of engagement, at the beginning, if you can, while you are still friends.

## Why now?

Why is the time ripe for you to start a new venture? Why is the time right for this particular venture?

It may be that the decision to start a new business is forced on you, if for example, you are made redundant from your job, or want to exploit a new idea before anyone else, or are just at that stage of your career. However, proper timing is important, especially from the standpoint of market development. Too late, and the market may already be saturated, with established brands. Too early, and the product may fail because of lack of need and infrastructure. Your product will need unique and compelling advantages in the market at the time of entry. If you can't list five such advantages, turn back now.

> TIME IS LIKE THE MEDLAR: IT HAS THE TRICK OF GOING ROTTEN BEFORE IT IS RIPE
>
> (CORNFORD, ACADEMICA MICROCOSMOGRAPHICA, 1908)

## Risk and reward

Entrepreneurship is not for everyone, and certainly not for those of a nervous disposition. Although the rewards are potentially great, only about ten per cent of start-ups last longer than five years. Even those that last five years, many end up as 'living dead' – struggling to make a living, but not actually failing. Only a few succeed and enrich their founders, often as much as a result of luck as of judgement.

> NOTHING, REPLIED THE ARTIST, WILL EVER BE ATTEMPTED, IF ALL POSSIBLE OBJECTIONS MUST BE FIRST OVERCOME
>
> (SAMUEL JOHNSON, RASSELAS, 1825)

Don't be put off by the fear of failure. The atmosphere has changed. It used to be the case that failure carried a large stigma, but nowadays almost the opposite is true, and many employers regard it as better to have tried and failed, than not to have tried at all. Some venture capitalists go further, and even require the principal players of a new venture to have one or two failures in their background, which they will have learnt from, and so be less likely to fail again.

> DON'T BE PUT OFF BY THE FEAR OF FAILURE

This is not to say that you should take crazy risks, or attempt the impossible. Setting up as an entrepreneur will affect not only yourself, but may have an impact on your family and close relations, and you need to consider and consult them as well. Think very carefully before you use the family house as security, even if banks seem reluctant to evict. There may be a significant opportunity cost involved in striking out on your own, for example you may miss promotion or other opportunities in your current employment.

None the less, it's a non-zero sum game. You might win lots, but if it all goes wrong will you be much worse off than you are now? You will still have the skills you have today, plus extra knowledge and experience that may make you even more employable.

It will, however, involve lots of time, stress, late nights and uncomfortable travel, and take a toll on your social life. You have been warned.

## Who needs it?

A particular failing of technologists is to get carried away with the technology. Technologists are in love with the technology. They build something because it can be built, rather than because there is a need for the product or service. The companies they start are typically technologically driven, rather than responding to market pull.

‘THE MOST SUCCESSFUL PRODUCTS AND SERVICES OFTEN HAPPEN ORGANICALLY, AS A RESULT OF CUSTOMER DEMAND’

The most successful products and services often happen organically, as a result of customer demand. A product that has been developed to solve one customer's problem can turn out to have a wider audience if lots of other people have the same or similar problem.

A technological advance may open a new market, but there must be a real or perceived need for the product or service.

## FAB: Features, advantages and benefits

You need to distinguish between the features, advantages and benefits of your proposed product or service.

*Features* describe the unique distinguishing characteristics of the product or service. For example, a feature might be:

- this program runs twice as fast as the previous version.

The *advantages* say why these features help. In our example the advantages might be:

- less waiting time
- uses fewer resources
- cheaper to run.

*Benefits* illustrate why these advantages are important, allowing you, for example, to:

- meet the budget
- experience less frustration
- get more done
- make your boss think you are better person.
- make you think you are a better person.

Benefits are often intangible, and may be perceived, rather than actual. Cosmetics, perfumes, clothes and cars are often sold on perceived benefits. 'Because I'm worth it' as the strap-line of one range of mass-market beauty products has it.

Technologists tend to emphasize the product features, but what actually sells are the benefits. This difference is one of mindset and technologists often find it hard to overcome.

# Types of business

## Product or service

Another factor is whether the primary income of the company is from products or services. There are major differences in company culture.

A product company does the same thing again and again. It needs to employ people, either directly or as sub-contractors, who are happy with repetitive work. Each product is only designed once, apart from occasional updates. Income is fixed by the amount sold and the profit on each. If you go for a cup of coffee in company time, it's reflected in the eventual profit.

A service company, such as a consultancy, does different things every day. It needs to employ people who are continuously creative. Income is on time sold. If you go for a cup of coffee, the client pays.

## Specialist or mass-market

Yet another differentiation is whether the company is specialist or mass-market. A specialist company may well know, or can identify all of its customers or potential customers. It may develop personal relationships with many of them. Feedback is direct, and the customers may work jointly with the company in developing new products or services.

A mass-market company, on the other hand, has a more constrained relationship with its customers, although it may have closer relationships with its distributors. Discovering what the market wants is harder, and the stakes are higher – get it right, and demand can exceed supply, get it wrong and the mistake can wreck the company.

## Brave New World, or Better, Faster, Cheaper?

Broadly speaking, new ventures can be divided into 'Brave New World' and 'Better, Faster, Cheaper' models. Both have problems.

Brave New World ventures set out to do something radically different from anything that has gone before. The difficulty with such ventures is that, because they are so new, there are many imponderables. Until something has been done, it is hard to tell whether the world wants it, how much they will pay for it or even what it is, and how much it will cost to develop. You're inventing the business as you go along.

On the other hand, Better, Faster, Cheaper ventures set out to do something that is already established, like sell books or wine, but aim to do it better faster and/or cheaper. Many internet ventures, like Amazon, fall into this category. The problem with such ventures is that there is only a limited window of opportunity before the incumbents, who may be sleeping giants but who are not totally dead, wake up and realize they can do it that way as well. Before this happens you need to have assembled your team, raised the finance, established your brand and grabbed enough of the market to have a chance of hanging on to some. You have to go like a proverbial rocket.

‘YOU HAVE TO GO LIKE A PROVERBIAL ROCKET’

Using the internet or better technology gives maybe a 20 per cent productivity advantage over a conventional business. Over the whole world economy that is a lot, but for a new individual business it is not that much. You had better be pretty good at whatever it is you are doing or propose to do, because just doing it on the internet is not going to give you enough competitive advantage to survive. The business has to stand on its own feet.

## Business models

‘What is the business model?’ is a polite way of saying ‘How are you going to make money out of this?”...

There are five serious (and one discredited) models for businesses:

1. Selling things for more than they cost.
2. Charging a commission.
3. Selling subscriptions.
4. Selling advertising.
5. Lotteries and other scams.

Of course, many businesses combine these models.

The discredited model says something like 'Hey, this is land-grab time. The market, in the absence of any other information, values us by the number of customers we have. We will do anything to boost our customer numbers, and then figure out later how to make money out of them.' Unfortunately, for most companies using this model, later never materializes, or at least does not materialize before the venture's funding runs out. This model is not new to the internet. Cable companies have been following it for years, as have companies in most bubbles – airlines in the 1920s, railways in the 19th century, and overseas colonization enterprises even earlier.

## Selling things for more than they cost

Selling goods and services for more than they cost is the fundamental model for most businesses. It assumes that:

- There is an accessible market for the goods or services at the required price.
- The competition cannot under-cut you in the chosen market.
- The stock required or capacity needed can be financed.

A special case is selling software or other intellectual property licences, such as chip design, computer games, or some drug developments. Here, although the cost of reproduction is low, the first one costs a lot to develop. This cost of development must be amortised over the number of units expected to be sold, and the company must have enough capital and resource to be able to get to first base and actually develop the product that it will eventually licence.

## Charging a commission

The second model is to charge a commission on what you, or other people sell. Online brokers, auction houses such as ebay, and business-to-business (b2b) markets use this model. They establish a marketplace, and hope people come to use it.

The issue here is liquidity: a market is only useful if you can be reasonably certain that you will meet the counter-party there, that is, that someone will buy what you have to offer, or conversely will sell you what you have come to buy. There is a real network effect: the more people there are in the market, the more useful it is. A competitor might try to split the market, but the network dynamics ensure that one market will dominate a sector.

## Selling subscriptions

In this case subscriptions are sold to gain access to a service. This is similar to selling the service on a per-use basis, but with an actuarial calculation to estimate the average use. Most internet service providers (ISPs) and telephone services are moving to a fixed-subscription, all-you-can-eat charge basis, such as BT's Surftime.

‘THE ISSUE HERE IS LIQUIDITY’

## Selling advertising

The model here is to subsidize the service you provide by selling advertising space, with the popularity of the service, such as a website or magazine, attracting the custom to the electronic equivalent of the billboard which you erect. The going rate for internet banner adverts (at the time of writing) is around £10 pcm, that is per thousand impressions. Given that the cost of employing a person is about £500 per day including overheads, the site needs at least 50,000 page requests per day, per person working on it, to break even, that is, to subsidize the cost of the service. This is tough, except for special cases like ISP homepages.

## Lotteries and other scams

There are various money-making models based on exploiting people with poor perceptions of the chance of unlikely events. The classic model is a lottery, which taxes the poor and the weak. A similar model is when the main source of revenue of a business does not come from the thing that is being sold, or even given away, but is generated from the means of accessing the product or service, such as a premium phone line or other access charges. For example, some TV shopping channels use hidden subsidy, deriving significant revenue from the premium phone lines that the customers need to use to bid for or to order the goods. Certain high-value TV quiz shows operate this way as well. The average cost of phoning a hotline to become a competitor on such shows is about £2 to the viewer, and hundreds of thousands of willing candidates phone each week. At the more dubious end of the scale is the straight scam of 'Phone this number to win a car', and other dodgy offerings.

## Exercises

**one** Write down five reasons why you should start your new business now; then write down five reasons why it is a bad time. Do the positives outweigh the negatives?

**two** Analyze your business idea in terms of the features, advantages and benefits of the new product or service. How will it make life better for those who buy it?

**three** Which of your personal goals will you satisfy by starting a business? For example, what is the relative importance of goals such as financial gain, desire for freedom, recognition, and realization of potential technology, to you?

**four** What are your present strengths and weaknesses in terms of experience, expertise, contacts and resources that will help you to launch your business?

## Further reading

Barrow, C. (1995) *Business Growth Workbook*. Kogan Page

Barrow, C. (2000) *Setting up and Managing Your Own Business*. Kensington West Productions

Barrow, C. (2001) *Business Plan Workbook*. Kogan Page (And other books by this popular and prolific author.)

Birley, S. and Muzycka, D. (1997) *Mastering Enterprise*. Pitman

Bygrave, W.D. (ed) (1997) *The Portable MBA in Entrepreneurship*. 2nd edn. John Wiley

Komisar, R. (2000) *The Monk and the Riddle*. Harvard Business School Press (Readable narrative about a business start-up in Silicon Valley.)

Raymond, E.S. (1999) *The Cathedral and the Bazaar*. O'Reilly (Inspirational book about the open software movement.)

Wickham, P. (1997) *Strategic Entrepreneurship*. Pitman

# 2 Developing the idea

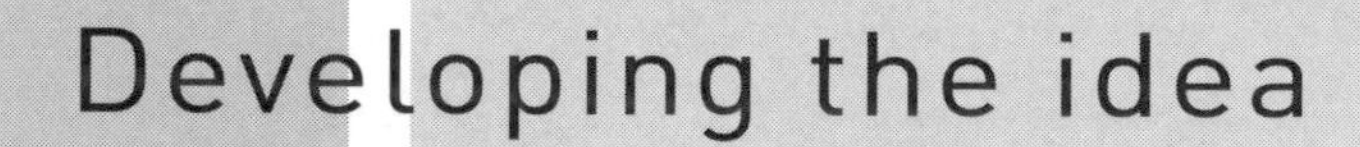

‘In my course I have known and, according to my measure, have cooperated with great men; and I have never yet seen any plan that has not been mended by the observations of those who were much inferior in understanding to the person who took the lead in the business.’

(Edmund Burke, 1729–97)

HOW DO YOU DEVELOP THIS GREAT NEW IDEA THAT will set you off on the path to fame and riches (or at least a second-hand Porsche)? At the beginning of this book we said 'so you've got an idea …' but turning an idea into a world-beating business proposition can often seem like winning a lottery. However, unlike the case with lottery tickets, there are a lot of things that you can do to load the odds in your favour, which, if done well, will significantly increase your chances of success.

It isn't just a question of 'good luck' or 'being in the right place at the right time' (although those things never did anyone any harm). Good ideas need to be well managed in order to meet their potential. The idea of a 'process' to help in the management of new ideas may sound out of place, but is in fact the best way to develop great new business ideas and take them forward to the marketplace.

Having a good idea is only part of the equation for success. There have been many wonderful ideas that have gone to waste because they were not developed properly. For example, academic research generates a wealth of great ideas, which until recently often stayed as just that: good ideas. Traditionally, the scope of much academic research did not focus on the application or exploitation of the idea and therefore good ideas rarely developed into new products or businesses, unless snapped up by external bodies. This situation is now changing with many academic institutions fostering close collaborations with industry, combining pure and applied research, and scholarly excellence, with real-world, practical experience. At the University of Cambridge, the new spirit of entrepreneurship has stimulated

> 'HAVING A GOOD IDEA IS ONLY PART OF THE EQUATION FOR SUCCESS'

the growth of the largest cluster of high-technology and knowledge-based businesses anywhere in Europe. This has had a phenomenal effect on both the local and national economies.

## The diversity of new ideas

First of all you need to understand that new product ideas are not all the same. Some are ground-breaking ideas, completely changing the way an industry works, although the vast majority consist of small improvements to existing products.

These different types of new products require a different marketing approach, and it is therefore important that you first understand what type of new product idea you are developing.

Breakthrough (new) products nearly always come about as a result of mixing different technology ideas. If you are hoping to develop a breakthrough idea, it will be very important to increase the breadth of experience and expertise both of yourself and of your 'ideas team'. This can be achieved through reading, research and networking.

- Read and research financial, economics and marketing magazines as well as the technical journals you are used to.
- Talk to people in different areas of expertise, or who do different things to yourself. Within an environment like a university, this could involve going along to lectures and meetings given by other departments. In an industrial environment, the equivalent is to read the trade journals of different industries, attend conferences for related industries or talk to people in parts of the business you don't usually meet.
- Think about how to challenge orthodoxy. Just because it has always been done that way doesn't mean that it is the right way to do it. Karl Popper in *The Logic of Scientific Discovery* points out how hard it is for new ideas to be accepted. History is full of people who have had to step beyond the norm and challenge orthodoxy to succeed, from John Harrison with his chronometers, via Charles Babbage with his

‘BREAKTHROUGH (NEW) PRODUCTS NEARLY ALWAYS COME ABOUT AS A RESULT OF MIXING DIFFERENT TECHNOLOGY IDEAS’

analytic engine as the first computer, to James Dyson with his cyclone vacuum cleaner as a modern example.

- See what they do in other places. Silicon Valley, just south of San Francisco in California, has for many years led the world in the IT sector, and what's hot in the US is likely to be so here in a year or two. Japan is innovative in consumer electronics and other places have centres of innovation or lifestyle elements that give a new perspective.

These suggestions don't guarantee success but they may help you to spot breakthrough opportunities and develop breakthrough ideas.

Success = Good idea + market application

When you have got your good idea, you will need to start thinking of market applications. Here is a checklist of how to go about developing your idea:

- Do some background research: collect information on the existing market and read up on other areas of expertise, not just your own.
- Work as a team: other people will help make your initial idea better.
- Work on your original idea: try to improve it and develop it into a 'solid concept'.
- Find out information about the market: how big is it and what do the customers in it want?
- Look to solve problems and meet needs that people have now.

Think through the risks in your new idea and work out how to manage your way to success.

Many of these points are addressed in a lot more detail at later stages of this book and will be needed again for the development of your new business. However, even at this early stage of generating ideas and developing a new business concept, it is important to be aware of how to test the viability of new ideas. Understanding what others will look for in your ideas is outlined in Chapter 4.

## Making it happen

Most really big new ideas come about through interdisciplinary collaboration, bringing together the knowledge of one area with that of others. Cross-disciplinary thinking requires excellent understanding and teamwork, and is often a very difficult process, which is why the products of such research are more likely to be breakthrough when they result. However, while new breakthrough ideas sound the most exciting, they can often seem like 'riding a tiger' and success is not guaranteed. You might get eaten.

It is very important to set boundaries for developing new ideas, or otherwise ideas can shoot off in any direction. First, think about what skills and experience you and your team already have (hopefully a 'multi-disciplinary' team) and consider the opportunities and limitations that these may present. This is very relevant if you are working for an existing company. However, if you are thinking of starting a brand new business venture, you will have few constraints or barriers to prevent you from going in whatever product direction offers you the best opportunity. That said, most successful new ventures are started by people working in, or close to, their own areas of expertise, just because they know most about that area.

SMART objectives:

- Specific
- Measurable
- Achievable
- Realistic
- Timely.

Think outside your box to get the best new ideas.

### Work as a team

Working as a team will be very important to you in your new business, starting right from the start. It is a common myth that great inventions are the work of a solitary inventor (often slightly mad) who does everything themselves. Such a scenario is very rare and most big inventions result from groups of people pooling their knowledge. If planned effectively, your team should encompass a wide base of experience and skills, optimizing your chances of perfecting that breakthrough idea. Teamwork is also important because it allows the workload and related responsibilities to be shared and gives a sense of joint purpose and encouragement. Above all, working as

‘IT IS A COMMON MYTH THAT GREAT INVENTIONS ARE THE WORK OF A SOLITARY INVENTOR’

part of a team can be fun and stimulating, and provides a support base to get you through the bad times, as well as people with whom you can celebrate your successes.

Teamwork will help in a number of ways:

- Add skills you don't have.
- See problems you don't notice.
- Spot new product formats you may have missed.
- Build things you can't construct.
- Give an unbiased view to 'your great idea'.

But now comes the rub. How do you get the potential from your team without giving away too much of the value? The bottom line is that you can't start a new business and keep all of the shares to yourself. Of course, this can work in your favour, as although you will have to share any successes, you will also not be left as the sole owner of a failure. The list of people who will share in your success will include anyone with a stake in the business, such as your investors and backers. However, by far the most important group will be your team, and it is essential that they feel adequately rewarded and appreciated for their contribution to making the idea work. Remember, you get richer by being a small fish in a big pond, than a big fish in a small goldfish bowl!

## Improve the original concept

There is another common myth that great ideas spring 'perfectly formed and developed' from the inventor's brain. This is rarely, if ever, the case, as most great inventions need a considerable amount of refinement to make them into world beaters. The sooner you start this process, the faster you will get it to market, although it is essential not to rush the development.

> ‘THE REFINEMENT PROCESS IS WHERE YOUR TEAM COMES IN’

The refinement process is where your team comes in. Give them the original concept and set them with the task of trying to improve it and fit it more closely to the needs of your target audience – the intended customers. (We are going to address market analysis and customer needs in the next section.) Your team will bring a range of different skills and perspectives to the original idea. It is very likely that, in the course of this refinement process, your team will come up with separate but related concepts. This can not only lead to other product ideas, but also enhance progress on refinement of the original idea, as working on more than one related idea at once can allow the best features from each to be combined.

It is important that you start to consider problems that may arise, or improvements that could be made, right from the start. Don't just wait for the problems to arise during development. Although refinements will be made along the way, you will waste valuable time and effort if you do not think through your options carefully from the start. This doesn't necessarily mean that you need to think of how to make the product more complex or sophisticated (especially as complex products often confuse customers, are costly to make and are more likely to have quality problems). In fact, the focus of your thinking should be how to make the product simpler and more targeted to the exact needs of the customer.

## Get information on your target market

One of the most common weaknesses in plans for new business ventures is that they have little or no information about the markets that they want to sell to. Unless you have a good knowledge of who might buy your product, and what their needs are, you will have great difficulty trying to refine it, and once finished it may not appeal to anyone, or meet anyone's requirements. Basically, you won't develop a world-beating new product and business if you don't know about your target market. More importantly, in the short term, you won't get anyone to invest in your business unless you can show them that you understand the market, what customers want and why you will beat your competition. Market demand is therefore integral to the success of the product, and you need to be thinking about it right from the beginning when you are working up your ideas and concepts. The key things you should consider are:

‘MARKET DEMAND IS INTEGRAL TO THE SUCCESS OF THE PRODUCT’

- How big is the market and how is it segmented?
- What problem/need are you trying to address?
- What functionality do customers want?
- What are the core benefits that you want to deliver?
- How will you beat the competition?

At a later stage in the development of your business you will need good solid answers to all these market and product questions, and more. However, these matters can't be ignored at the point of idea generation and concept development and they are critical to the work of improving your initial idea discussed in the section above. We discuss marketing in more detail in Chapters 13 and 14.

## Identify your customers' needs

As you are working on your wonderful new business idea, killer application or fantastic invention, it is important not to lose sight of the market that your product is aimed at, or what your customers will want. Try to understand not only the size and potential of the market but also what your customers are like and how you can best satisfy their needs and desires. How do your potential customers' needs match with the 'functionality' of your product, i.e. what it does?

A good method of considering this is to draw a simple table with two columns, describing on one side the functionality of your new product, and on the other, the needs of the customer. Now try to match up the key product features with the key customer needs and then rank them. How well do the sides of your table match? Do you meet all the key needs of your customers? Are you focusing your work on the features that match with your customers' needs? If not, you should be! A problem looking for a solution is *much* better than a solution looking for a problem.

' A PROBLEM LOOKING FOR A SOLUTON IS *MUCH* BETTER THAN A SOLUTION LOOKING FOR A PROBLEM '

## A brief introduction to market analysis

So, where do you find information about your target market and potential customers? Well, it will require some hard work from you and your team but fortunately, such information has never been so easily available.

Here are some types and sources of publicly available data:

- internet searches on markets, competitive companies, journal articles, etc.
- conference papers, published market surveys, government statistics, OECD publications, *FT* info, *Economist* reports, *Business Week* surveys, Bloomberg data, Reuters information, trade association reports, specialist trade journals
- product information, brochures and reports from competitors.

### Desk research

The first place to start with any market research project is to do desk research, that is, researching information you can gather by sitting at your desk, such as information on the web, in other media or in the local library. You can discover what the existing market or solutions are, and what the competition is, both actual and potential. This can be done by looking at company websites and annual reports, from which you can deduce a company's sales figures, and whether they are profitable. You can compare your proposal to what you find out, and then ascertain any unique features that your product has, and think about who they might appeal to.

‘YOU ALSO NEED TO LOOK AT THE DEMOGRAPHICS’

You also need to look at the demographics. Just how many people are there who would buy your product? For example, if you have a product that would sell to, say, schools, then how many schools are there in the country? What are their key motivators in relation to your product type? How much disposable budget do they have? How do other people who sell to the same demographic operate?

## Where do you fit?

Some people use 'bins' or psychological profiles. Traditionally there are two genders, seven age ranges and five socio-economic classes (A, B1, B2, C, D), making some 70 categories. You can buy fairly detailed descriptions of each category from market research organizations, including details of their typical buying habits, disposable income, leisure pursuits, etc. You can even get lists of names by classification, or conversely buy databases that classify the likely category of the residents of each postal code.

Other marketing strategies distinguish categories of customers by naming lifestyle groups, such as Sinkys and Dinkys (Single Income No Kids Yet, and Dual Income No Kids Yet), or Empty-Nesters, people whose kids have left home. This can be useful as such categories are likely to have above-average disposable income. The fastest-growing demographic for the internet is women over 55, possibly a reflection of a category of people with disposable time.

The UK government's national statistics use a classification based on occupation, with currently nine major groups. Details are given on their website **www.statistics.gov.uk/nsbase/methods_quality/ns_sec/soc2000.asp**:

1. Managers and senior officials
2. Professional occupations
3. Associate professional and technical occupations
4. Administrative and secretarial occupations
5. Skilled trades occupations
6. Personal service occupations
7. Sales and customer service occupations
8. Process, plant and machine operatives
9. Elementary occupations.

It is often helpful to write a pen-portrait or two of a typical customer, to focus the mind.

Desk research can also suggest suitable distribution channels and distributors, and where to place the advertising or other marketing communications. It can suggest what the target audience might read or watch, and whether there are specialist magazines that cover the area.

## Market surveys

Market surveys involve actually going out and asking potential or real customers what they think. If you think you will sell 10,000 units, then find ten of the sort of people who might buy and ask them what they think. If you can't even find those ten, then you are already in trouble.

‘GETTING RELIABLE RESULTS FROM MARKET SURVEYS IS AN ART AND A SCIENCE’

Getting reliable results from market surveys is an art and a science, as the answers you get can depend on the questions you ask. It is best left to professionals.

Surveys can be divided into qualitative and quantitative categories.

**Qualitative surveys** Qualitative survey techniques find out soft information, like desirable features or ease of use, or what people remember about an advertisement, without putting hard numbers against them. They include techniques such as focus groups and usability testing, where people are observed using the product to do some task. Qualitative methods are great for gaining insight into what is important, and the results can then be followed up with a quantitative survey to give the actual numbers. Qualitative methods also tend to be quite small scale, and hence cheap compared to a major quantitative survey, for example:

- Talk to some key customers, users, distributors or agents.
- Talk to your partner, friends and relatives (you would be surprised how much ‘product use’ insight you can get this way).

- Talk to colleagues or other people you know who have worked in this industry area (or related ones).
- Organize focus groups (these are really quite easy to run and can give you very good insight into how people may consider a new product). The group should comprise of potential future customers – probably between six to ten in number. Organize an informal meeting for the group to come and talk about whether they would be interested in your product and how they might use it if they were. A few bottles of wine can help to lubricate this process, and shake off any inhibitions! The amount of data that can be collected and the insight that such groups can give is often quite surprising.

**Quantitative surveys** Quantitative methods generate numbers, usually by surveying a sample of the target population, and then extrapolating to the full population. Reliable results require careful statistical design. Pricing is particularly hard to ascertain, and may need correlation between several surveys asking about different feature combinations.

Surveys can be face to face, for example by those irritating pollsters who stop one in the street, or by post, or by telephone. Postal questionnaires can only expect a return rate of only a few per cent, even with the offer of an incentive gift or payment. Telephone surveys have a higher hit rate, typically a bit under 50 per cent, although the rate depends on the exact definition of what is meant by acceptable response.

Quantitative methods also include test marketing, where the product is marketed in some limited locale or to a small target group, and the results scaled to predict the response of the whole country or larger market. While test marketing is the most accurate way of assessing market size, it takes a long time, and few products have an opportunity window long enough to allow a full market trial before launch. If the market trial is unsuccessful and the product not launched, then there may be ongoing support problems for the few that were sold.

## Knowing when to start

There comes a point when you have talked through your new business idea, you have built a team of energetic, enthusiastic people, you have worked on the original idea and made it better, you have collected some market information, and have a really good idea of why people will want your new product.

This is the most critical point in the life of a new business, and it is at this juncture that you need to make the decision of whether to take your faith in your hands and commit to starting the project in earnest. The biggest single cause of good ideas failing to make it to market is because the team bottle out at this crucial stage, and don't have the courage of their convictions to actually make a go of it. A common reason for why so many teams lose their nerve is because they are not prepared to take the risk. Indeed, how *can* you be sure? How do you know that you have a winner? How can you ask your friends to trust you with their savings, their time, their futures, if you are not sure yourself?

### Assessing the risk

There is always a certain amount of risk involved in starting a new business, as there is in everything we do! However, if the risks are assessed sensibly, and managed well, you should be able to make an informed decision as to whether to launch your business or not. Your answers to the following questions will give you some good pointers as to the levels of risk involved, and whether these risks are worth taking:

- Does your new idea excite your team?
- Does it convince other people you describe it to? (Including partners, friends and relatives – they may have more sense than you think!)
- Do you understand the market that the product will address and the customer needs that it will meet?
- Do you think it is a better product than competitor products? (Failed internet start-ups have reminded us all that it's no good being 'another company' – if you have no clear advantage, your business will not work.)

- Do you have a credible team with many of the skills you will need to start your new venture? (Not only the technical skills – what about the business skills as well?)
- Do you have any protection for your new idea – copyright, patent, registered design, etc.?
- Does your new product have a good fit with the 'timing' of market growth, changes in customer needs, regulatory requirements? These are all good things to try to fit in with.
- Are you ready to start working on developing the key issues for your business plan? This is the detailed document that will describe how you expect the business to work; it will also be the document you will use to 'sell' your new business concept to investors.
- Are you and the rest of your team ready for one of the most exciting times of your life, but one that will also take over 24 hours a day of your time and all your energy for the next two years minimum?

However, let's just do a quick review of the actual risks that may affect the chances of success for your new business venture. Risk is perceived as a tricky concept and is often not very well understood. The temptation with difficult things like this can be to ignore them and hope they will go away, which in this case could be dangerous.

Estimating risk for a large project can be very difficult, but if the risks are broken down into smaller individual areas, it is possible to make quite good estimates of risk. We can also work out the best ways to approach the key areas of risk so we don't jeopardize the whole project. Here are typical main categories of risk to consider:

“ESTIMATING RISK FOR A LARGE PROJECT CAN BE VERY DIFFICULT”

- technical risk
- market risk
- commercial risk
- team/people risk
- financial risk

- manufacturing/production risk
- collaborator risk.

It is important that you consider all of these areas. Many technology-based teams focus uniquely on the technology risk whereas, in fact, this is often more easy to assess and manage than some of the other areas of risk.

With a new business venture, it is worthwhile setting aside a few hours with your team to review the key risks in the business and to identify which are the most important and critical to success. Don't avoid risks or sweep them under the carpet, but instead plan how to address and manage high-risk or key success areas, as early as possible in the project. This will save you, your team and your investors a great deal of time and money.

“DON'T AVOID RISKS OR SWEEP THEM UNDER THE CARPET”

Risk has a number of elements. Just because some event could occur, it doesn't mean it will, or that precautions are cost-effective. You need to assess the probability of an event happening, and the damage which will occur if it does. Mathematically:

$$\text{Risk} = \text{threat} \times \text{vulnerability} \times \text{value}$$

where impact is typically measured as the value of the asset that you are protecting. A risk analysis method might be:

- For each risk score the likelihood of the threat occuring as a probability. You might just want to characterize it as low, medium, or high, say 0.001, 0.1, 0.75 probability.
- For each risk score the likelihood of the threat succeeding if it occurred, also as a probability.
- Score the value if it does succeed, for example in financial loss to the company.
- Multiply the likelihood of occurrence by the chance of it succeeding and by the value of the impact, and rank risks in order.

This exercise helps to concentrate on the important risks.

You can now assess the cost of the counter-measures, either to prevent the occurrence of the risk, or more likely, to reduce the probability and limit the amount of damage which could be caused, to the point where the risk can be economically insured against.

Don't be put off if your first risk assessment makes your exciting project look scary and very dodgy – all new businesses start that way, you are just being more realistic than most people starting up. Don't lose heart, but instead work out with your team how to address the key areas of risk, and then, if you are still all convinced to give it a go, CONGRATULATIONS!, you are ready to make a start with your new business venture.

‘DON'T BE PUT OFF IF YOUR FIRST RISK ASSESSMENT MAKES YOUR EXCITING PROJECT LOOK SCARY AND VERY DODGY’

## Developing a plan

Planning is not very glamorous and nobody really likes doing it. Now that you've decided that you are going to start this new business, you will probably be itching to just get on with it, get into the lab, do some marketing, meet with venture capitalists, talk to potential customers …

‘TIME SPENT ON RECONNAISSANCE IS SELDOM WASTED’

(FIELD-MARSHAL MONTGOMERY)

WAIT. Although it may sound boring, the most important thing for you to do at this point is to get together with your team and spend a day or so planning the key issues that you need to address. A key component of this is writing a project plan, making sure everyone is clear about who is going to do what, and ascertaining that you have a realistic level of resources to achieve success. Be sure to identify milestone and decision points on your project plan, at which you can review progress. These will help to make sure that you are going in the right direction and making adequate progress. The project plan should cover all of the key points that you are going to write up in your business plan and must include:

- technical development programme (including protection of any intellectual property)
- marketing plan
- production plans

- financial analysis 1 – how much money you need
- financial analysis 2 – where you can go to get the money you need and what the sensible milestones/funding phases are
- plans for team building, recruitment and resources, and organizational plan
- details of ownership, equity and how these will change during the funding rounds and with further recruitment
- identification of key collaborators, partners, agents, etc.

Write the plan up and also develop a Gantt chart (see Chapter 10). Make sure that you review progress against your plan at regular intervals. When the project has gone completely off-plan (don't worry, it will), don't just tear it up and ignore it, but use it to help you write up a new plan. Ask yourself why you went off-plan, and what can be done to remedy this? By refining your plans, you will be able to keep track of reality – without this approach the project will wander along in a very haphazard fashion, going up blind alleys, getting lost and, more crucially, losing everyone a lot of time and money.

In summary, develop a plan with the team; write it up; use it and review it; modify it; make it work for you. You won't regret it.

## Ideas of the future: some emerging areas for new businesses

This section contains speculations on where ideas for new businesses might come from.

We live in a time of unprecedented technological and social change. This environment can give rise to many difficulties but also presents outstanding opportunities for new enterprises. Opportunities for new ventures come at the raw edges of change.

Change is rapid. Moore's law, applicable to change in the world of computer technology (**http://info.astrian.net/jargon/terms/m/Moore_s_Law.html**), predicts a factor of change of two every 18 months in almost any measurable parameter, such as computer processor speed, memory size or even amount of documentation. Much to everyone's surprise, this law has held true ever since computers were invented, and shows no sign of stopping. However, most development is quite slow. Innovations do not come out of nowhere, but are based on years of research and development. The time taken for an idea on a lab bench to develop into a widely accepted mass-market product is typically about ten years. By closely monitoring what is coming down the development pipe you can start to predict developments in technology and markets in the medium term.

It is also useful to look at some of the forces driving change. For instance, in recent years the dramatic increase in computer power has had a powerful influence on new technology. However, the related reduction in communications costs has had an even greater effect. The change from copper to optic fibre brings about, in principle, a reduction in the cost of bandwidth by a factor of 100,000. When you consider that the Industrial Revolution was triggered by a reduction in the cost of energy by a factor of a few hundred, the recent reduction in communications costs can be appreciated as a major driver of change. Indeed, this change is as profound as the invention of the printing press.

“THIS CHANGE IS AS PROFOUND AS THE INVENTION OF THE PRINTING PRESS”

Imagine a world of universal high-speed communications. What will the communications be used for? What social changes will they bring? While books (e.g. Toffler, *Future Shock*) can, and have been written about these topics, a few pointers are raised here.

## Uses for bandwidth

We have already seen the content of the web change from being mostly text based to predominantly picture based. The next major step is the development of more moving pictures and multimedia content, leading to the development of three-dimensional moving images.

Greater bandwidth is leading to a convergence of communications (such as the telephone), with computing and entertainment (such as the TV). Digital television technology is enabling people to interact with the family TV to make purchases, join chat groups, access the internet, or send e-mail. Computer games now gross more than movies, and have penetrated more than 75 per cent of households. Such developments are changing the nature of entertainment.

Universal communication allows us to pull content from the internet cloud originating anywhere in the world, and is breaking down national barriers and creating an increasingly global market. At the time of writing (spring 2001) there are already some 200 web-based TV stations, and some 5,000 web radio stations, with more starting up every day. Soon a web user will be able to pick up any broadcast from anywhere in the world. I'm sitting in Cambridge UK listening to KDFC San Francisco as I write. For some sorts of music (such as specialist areas like French chansons), mp3s are already the dominant publishing model.

Greater bandwidth also enables data streaming such as continuous news feeds, or multimedia and online video presentations, which are changing the web model from mostly pull, where the user controls the flow of information and pulls the content as needed, as with most web pages, to mostly push, where the originator controls the content and pushes it down the pipe to the user, as is the case with a broadcast TV station. This sea change within the web model has serious consequences for the economic model, and consumer choice. For example, the inserted advertisements on the web, as on TV, are chosen by the originator and are unavoidable by the user.

On the other hand, development of anonymous file copy programs, such as Napster and Freenet, change the role of digital copyright and the enforceable rights of authors. They also change the emphasis from routing-based addresses, such as conventional URLs, to content-based access, where the content is addressed by what it contains, rather than where it is stored. This change reinforces the role of search engines and intelligent agents.

Communication becomes universally pervasive. The internet regards censorship as damaging, and routes around it. The mobile phone in your pocket is your gateway to the infosphere, carrying your electronic identity. Developments such as the Microsoft. NET infrastructure allow ubiquity, but also enrich the gatekeepers, that is the mechanisms which you have to pass through to gain entry to the information that you want, either paying a fee or leaving your valuable details.

## Social consequences

Barriers to publication are dramatically lowered through the influx of new forms of communication such as the internet. Anyone can publish, regardless of the merit of the material, and, more importantly, regardless of local customs or approval from society or official organizations such as governments. The worldwide knowledge base, encompassing both 'good' and 'bad' stuff is available to anyone with access to the internet. Some commentators have predicted that the empowering effect of access to such a huge information bank will create a social divide between those with access and those without. However, many believe this to be less significant than, say, the divide between the literate and those who cannot read, or the numerate and those who can't add up.

Another social consequence of new technology is that it is becoming less important where you live. Communities are formed based on interest, rather than on geographical location. You want to talk to the other few hundred people that share your passion for lesser Etruscan pottery? No problem – there are something like 90,000 specialist public mailing lists and about the same number of newsgroups for every subject under the sun, and then some. You want to have an open video tunnel between your desk and your colleagues, on the other side of the world? Easy, just a few clicks.

> ‘ANOTHER SOCIAL CONSEQUENCE OF NEW TECHNOLOGY IS THAT IT IS BECOMING LESS IMPORTANT WHERE YOU LIVE’

## Technical consolidation

The technical consolidation of products after they leave the lab and become part of everyday life is very slow. At a critical point in this process, formal standards evolve which allow multiple manufacturers to make compatible products, and allow a market to develop that benefits all.

As standards evolve, there is scope for more product innovation and for new industries to exploit the gap between the technical development and the saturation of the market.

Examples of emerging standards include the various mobile phone standards, digital television standards, internet standards such as IPv6, secure network transaction standards, new classes of drug protocols and so on. Monitoring the bodies which form such standards, for example the Internet Engineering Task Force (IETF), or DVB (Digital Video Broadcasting), can give glimpses of the future.

## Communication, not content

It turns out that communication, not content, is king. An analysis of any ISP's logs will quickly show that the majority of use is from people meeting and chatting to other people online. People love to talk. Even fast-growing specialist areas, such as genealogy sites, are ways of meeting people, in this case possibly distant relatives, and sharing interests. Services that put people in touch with each other are good places to look for new business opportunities.

‘COMMUNICATION, NOT CONTENT, IS KING’

## The trillion-dollar market

Commerce on the internet is another area for new ventures. The internet is a trillion-dollar market opportunity. Some argue that by using the internet, productivity gains of around 20 per cent can be made. Such savings are typically made through a reduction in the cost of sales and marketing, but can also be as a result of the increased efficiency of the market. One particular area of increased efficiency could be through the use of electronic information to enable just-in-time delivery, with consequent lowering of inventory cost. Across the whole economy this efficiency amounts to over a trillion dollars worth of savings, which is more than enough to pay for the infrastructure required.

Enterprise migrates to where labour is cheapest, and the infrastructure and regulation friendliest. However it owes no allegiance, and can move somewhere else just as easily. Jurisdiction is still an unsolved issue, but an internet transaction is probably governed by the rules and nationality of the banking system and credit or debit card processor.

## Death of the nation

Electronic commerce allows the development of global markets, and global companies. The nation state is dead, replaced by smaller localities, and wider issues. Do I care that the customer care centre is in India, the credit card company in Bermuda, and the factory in China? So long as the process is seamless, the effect is that they are all no further away than my screen. This can have dramatic effects on national economies, for example on tax revenue collection, and is further breaking down national barriers.

I don't care, or even know, where the people that share my interests online are located. I don't care where I purchase goods and services from, just so long as they are efficient suppliers – I pull what I need from the internet cloud. My stockbroker could be located on the moon, as long as they look after my money properly. All these things are accessed via my screen.

> I DON'T CARE, OR EVEN KNOW, WHERE THE PEOPLE THAT SHARE MY INTERESTS ONLINE ARE LOCATED

I do, however, care about the people I meet in my local pub, school PTA and very local community, but my interests, and more importantly, where I pay my taxes, are effectively either very local or are global.

All this will have a profound effect on the way communities are organized. Already we are seeing moves towards regionalization, with for example, regions such as Scotland or Wales receiving direct grants form the supranational European community, and in turn bidding for multi-national companies to set up in the region. This trend will probably continue, and even accelerate, with national governments looking increasingly irrelevant. Already private companies can and do provide much of what was formerly the prerogative of national government, such as healthcare, pensions, education, security and policing, transport, sewage provision, water, energy, telephone and other utility supplies, and this gives opportunities for new ventures that did not exist a few years ago.

Who will be the main employers, and the major industries in this future? Just as agriculture now employs a small fraction of the number of labourers it did a century ago, yet has greater output, so the manufacturing industry is rapidly shedding people. Some predict that in the future there will only be three industries with any significant numbers of employees, and these are education, healthcare and entertainment in its broadest sense. If you want to prosper, set up a company with global potential in one of these areas.

## Exercises

**one** Write a page of predictions. File it somewhere safe, and look at it in a year. Update it again next year.

**two** Brainstorm. Write down ten ideas for a new business, no matter how crazy or impractical. Develop criteria for selection, such as benefits for the user, practicality, existing competition and fun to do. Score your ideas against your criteria. Select the best and develop five reasons why this is a killer idea.

## Further reading

Popper, K.R. (1992) *The Logic of Scientific Discovery*. Rev. edn. Routledge

Toffler, A. (1985) *Future Shock*. Pan

### *Internet demographics*

**www.mids.org/weather/** (Internet traffic congestion (measured by lag time) presented as an animated 'weather' map.)

**http://cyberatlas.internet.com/** (US-based market research about the internet.)

**www.netcraft.com/survey/** (Web server software usage on internet-connected computers. Tells the number of server computers connected to the internet, from a friendly UK company.)

**www.lib.umich.edu/libhome/Documents.center/stdemog.html** (Survey-based statistics about the web for University of Michigan Document Center.)

# 3 Protecting the idea

‘As the births of living creatures at first are ill-shapen, so are all innovations, which are the births of time.’

(Francis Bacon, 1561–1626)

IPR STANDS FOR INTELLECTUAL PROPERTY RIGHTS. They are your rights in your idea. These include formal rights such as:

- patents
- copyright
- trademarks
- registered designs.

The term can also cover informal rights such as trade secrets, customer and contact lists, distribution licences, and the complexities involved in large software programs that may be only understood by one or two key staff. Very often, such informal IPR is the most useful, as any other involves some form of disclosure. Protecting the IPR does not always protect the idea. The best way to keep control of your idea is to not tell anybody who does not absolutely need to know about it, and if you do need to tell someone, make sure that you contractually bind them to keep the secret with a non-disclosure agreement (NDA). An example of an NDA is given at the end of this chapter. However, secrecy is sometimes not practicable, or compatible with the business, and other more formal methods, such as patents need to be used.

‘PROTECTING THE IPR DOES NOT ALWAYS PROTECT THE IDEA’

IPR issues have particular relevance for software- and internet-based companies, where the intellectual property may be the greater part of the asset value of the company.

Even if you have formal protection in place you cannot relax. Suppose a major corporation with bottomless pockets brings out a product similar to yours. What are you going to do? Sue? It might be very hard (and expensive) to prove in court against hostile opposition that they have indeed copied your product or infringed your patents. Even if you could, you may not be able to afford the legal fees, and wait the time it takes. A better response might be to exploit the fact that small companies are quicker on their feet than large ones, and have version 2 ready to roll, with the extra features, while they are still developing Version 1.

The subject of IPR and its detailed regulations is very complex, and at best, it is only possible to present a brief overview here. This is not legal advice, and should not be relied upon as authoritative, and no responsibility can be taken for errors, omissions or loss suffered. You are strongly advised to get professional help when dealing with such issues.

## Patents

A patent grants an absolute right to a monopoly on an invention for a limited period of time, usually 20 years. Even if someone else independently invents the same thing, the holder of the patent can stop other parties from exploiting the idea for the period of the patent.

For an idea to be patentable it must be novel, which means that no one can have published it, or made it available to the public, for example incorporated into a product before the filing date of the patent. It cannot be an idea that has been published or well known previously ('prior art'), even by the people applying for the patent.

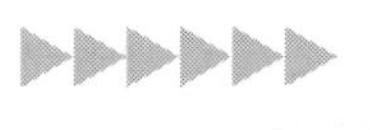

Note: This can be a particular problem for research results, such as a PhD thesis, and it is important that any patents are applied for before the thesis is submitted, or, any research results are submitted as a learned paper. Ensure that you have a confidentiality or non-disclosure contractual relationship with anyone you show or discuss your idea with prior to applying for the patent.

It must be a non-obvious invention, and not just an idea moved from one format to another. It must also be practicable, and capable of being built from the description by 'one skilled in the art'. In the US it must be useful as well. Thus, you cannot patent a perpetual motion machine, or a ladder to the moon made of eggshells.

What can and cannot be patented is currently in a state of transition. Up until recently, only actual devices and technological methods or processes could be patented. Mathematical formulae, genes and software could not. However, this situation is changing rapidly, and a global debate has been fuelled regarding whether software and ways of doing business should be eligible for protection by patents. At present (2001), such things are generally protectable in the United States, whereas in Europe, software can only be patented if it gives rise to a 'technological effect', and pure business methods cannot be patented at all. Progress is slowly being made to change this situation, although UK law has not yet been amended. In practice, software often can be patented in the UK and Europe, as can certain biological processes and gene sequences, although non-technological business methods, 'look and feel', and mathematical formulae cannot.

‘WHAT CAN AND CANNOT BE PATENTED IS CURRENTLY IN A STATE OF TRANSITION’

In the US, software must be something that is reducible to hardware. Thus, software on its own cannot be patented, but a memory chip or disk containing the software can be. In general, 'look and feel', or a business method such as Amazon's one-click ordering, can be patented, and there is even a patent on a method to boil rice! Some recent method patents (including the one-click patent) have been challenged on appeal in the courts, and, although at the time of writing there is no final decision, it would appear that the more obvious method-only patents will not be upheld.

Patents are very powerful as they confer an absolute monopoly. However, they are also very complex and, as legal documents, are written in a particular style and format. Patents require expert handling and it is strongly advised that you consult a suitable patent agent to deal with them.

The whole process is expensive and time consuming, taking up to four years or even more to get a patent granted, and costing, with professional fees, tens of thousands of pounds if patented in more than one country. Even then you are not home and dry, as a patent can still be challenged in court, and it is also up to you to detect and prosecute infringements, whether it is unauthorized people making, selling, importing or even using your patent within the protected jurisdiction. Renewal fees are also payable annually, and failure to pay them makes the patent invalid.

Having got your UK patent, it will only provide you with protection in the UK, and if you require protection elsewhere, you must apply to separately in each country and jurisdiction. However, there are now common European and worldwide patent offices, and it is possible to apply them for a patent that applies to all EU countries in the case of the European office or to all those countries that recognize the Worldwide Patent Organization (WPO) in the case of the worldwide offices. International agreements recognize the date of first filing in any country as the priority date. If you are going to apply for an international patent, you must do so within 12 months of your first application in the UK (or your own jurisdiction), or you don't get the priority date. It is, however, usual to apply for a local patent first, as a local search is much easier and cheaper. Fees may be payable for each jurisdiction in which you wish the patent to be valid, which can soon become very expensive.

‘SOMETIMES IT IS BETTER JUST TO MAINTAIN SECRECY’

One disadvantage of patents is that, once a patent is published, it becomes public, and therefore anyone, including your competitors, can read it. Sometimes it is better just to maintain secrecy.

## Copyright

Copyright is exactly what it says: it is a right which prohibits copying. It does not grant a monopoly, nor prevent independent invention. It cannot be used for names or phrases, but can be used for complete designs. Copyright is widely used as protection for software. Copyright does not protect the idea, only its expression. Indeed, some

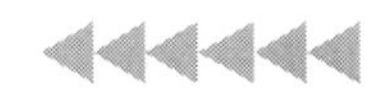



The following shows the normal procedure.

This table sets out the basic steps necessary for your application to progress to the grant of a patent. It explains, for each stage, what action you need to take and what action the Patent Office takes. A fast-track service is also available, which is referred to as *accelerated processing*.

| What you need to do | When | Form/fee | What the Patent Office does |
|---|---|---|---|
| File a full *description* of your invention; you can't add anything later. | Before you tell anyone about your invention or use it in public. | Form 1/77 | Gives your application a number and a filing date. Writes to tell you what you need to do next. |
| File *claims* and an *abstract*. | Within 12 months of your filing date. | | Adds the claims and abstract to your application. |
| Request a *search*. | Within 12 months of your filing date. | Form 9/77 (£130) | Issues a preliminary examination and search report within about 12 weeks. |
| After 18 months – Publishes your application – 'A' publication. | | | |
| Request examination. | Within 6 months of 'A' publication. | Form 10/77 (£70) | In date order – issues a substantive examination report. |
| Reply to examination report. | Within the time specified in the examination report. | | Looks again at your application with any amendments you have proposed. |

The above step may be repeated until the final form of your application is agreed. Then the Patent Office will – within 4 years from your filing date – grant you a patent and publishe your application in its granted form – the 'B' publication – and then issue a Certificate of Grant.

software systems have been deliberately re-implemented as 'clean-room copies' by teams with no knowledge of, or access to the original code.

Copyright is self-declarative. The author automatically owns the copyright of the work, although may be assigned, for example, to an employer automatically as part of an employment contract. At one time, it was considered necessary to include the word 'Copyright' (or the copyright symbol ©) together with the year of publication and the name of the author on the material. However, this is no longer strictly required, but does create a presumption of ownership. The presumption can, however still be challenged and refuted. A typical copyright statement might be:

Copyright 2000 Jack Lang

A statement of rights, stating explicitly what the user can or cannot do with the material can conveniently follow the copyright statement. For example:

'Copyright © 1989, 1991 Free Software Foundation, Inc. 675 Mass Ave, Cambridge, MA 02139, USA. Everyone is permitted to copy and distribute verbatim copies of this document, but changing it is not allowed.'

Copyright for new work lasts for 70 years after the death of the author, although some jurisdictions vary this. Old works are protected for differing lengths of time, dependent on the laws prevailing at the time that they were copyrighted. Under the Berne Convention, which was signed by 96 countries, copyright is recognized internationally, although standards of enforcement vary.

Copyright is incremental. For example the translation of a work carries both the original copyright, and the copyright of the translator. Aggregations, such as a database, may be copyright in their own right, as well as the original copyright on the individual entries. There is also an entirely separate European 'database right', analogous to copyright but for databases and other collections of information. The individual items of information may not be original, but the database taken as a whole is an original work.

'Fair use' of copyrighted material is permitted. For example, limited use of works may be possible for research and private study, criticism or review, reporting current events, judicial proceedings, teaching in schools and other educational establishments and not-for-profit playing of sound recordings. Fair use normally only applies to text, although hyperlinks seem to be OK. There is a limit to fair use of copyrighted text, which is normally something like 300 words (about a paragraph).

Detection and prosecution of copyright abuse is the responsibility of the copyright owner, although organizations such as Trading Standards and Customs and Excise can help. FAST (Federation against Software Theft) is doing good work in this field and should be supported.

“DETECTION AND PROSECUTION OF COPYRIGHT ABUSE IS THE RESPONSIBILITY OF THE COPYRIGHT OWNER”

It can be quite difficult to prosecute infringements of copyrights, as direct copying must be shown, and you will need to prove that the material copied is yours. One way to ensure against this problem with words in machine-readable form is to include steganographic (hidden) signatures in your original. For example, in a software program, you could include harmless but nonsensical code, which would be unlikely to be included in any re-invention. The presence of the sequence in another program would indicate that copying has occurred. However, with such a high incidence of copyright infringement, the issue is often not so much whether the infringement has occurred, but whether prosecution is feasible or economic.

## Copyright and the internet

Copyright has become even more complicated with the advent of the internet. The problem is that copying is so easy – you just need to point and click.

The boundaries of what constitutes illegal copying on the internet are blurred. Forwarding an e-mail to a friend is probably OK, but publishing an e-mail on a website is probably not, unless prior notice has been given, such as for an automatic archive for a mailing list, notified when you join the list.

A link to a website can probably be considered as fair use, even with a short textual extract, such as those used by a search engine. However, if a graphic (or even a thumbnail of the graphic) is added, the fair use conventions are violated. Repurpose the other site by displaying a page of it inside your frame, and you may have violated the copyright, especially if the authorship is not acknowledged.

The case law usually quoted is the *Shetland Times* case. Briefly, the *Shetland Times* published a website, with the front page containing headlines and links to the web pages containing the story. The *Shetland News*, a rival publication, published a page containing copies of some of the headlines and links, but without acknowledgement, thus bypassing the *Times*'s front page and advertising. This was held to be a breach of copyright, although the matter was eventually settled out of court, so no formal precedent has been set.

## GPL and open source

An interesting use of copyright is that pioneered by the Free Software Foundation (**www.fsf.org/fsf**). They use copyright in the opposite sense to usual, to ensure that when a document such as a piece of software is copied, the copies can then also be freely copied, as can any modifications or developments.

The Free Software Foundation has developed the General Public Licence (GPL) that explicitly permits such copying, provided that the document is copied in its entirety, with the copyright notice and conditions attached. Note that although the document can be distributed, its uses are covered by the GPL licence, and, although derivative works can be commercially exploited, the original cannot be.

GPL and its derivatives have become the default choice for the distribution of source code, and are used, for example, by Linux, Mozilla and Freenet, providing a useful and powerful mechanism to enable collaborative software development by groups and people not otherwise related.

## New legislation

In the US, on 28 October 1998, President Clinton signed a revolutionary new piece of legislation into US law called the Digital Millennium Copyright Act. According to the *The UCLA Online Institute for Cyberspace Law and Policy* (University of California at Los Angeles) the main implications of this legislation are that it:

- Makes it a crime to circumvent anti-piracy measures built into most commercial software.
- Outlaws the manufacture, sale, or distribution of code-cracking devices used to illegally copy software.
- Permits the cracking of copyright protection devices but only to conduct encryption research, to assess product interoperability and test computer security systems.
- Provides exemptions from anti-circumvention provisions for non-profit libraries, archives and educational institutions under certain circumstances.
- In general, protects internet service providers from copyright infringement liability for simply transmitting information over the Internet. Service providers, however, are expected to remove material from user's websites that appears to constitute copyright infringement.
- Limits liability of non-profit institutions of higher education, for instance when they serve as online service providers, and under other certain circumstances such as copyright infringement by faculty members or graduate students.
- Requires that 'webcasters' pay licensing fees to record companies.
- Requires that the Register of Copyrights, after consultation with relevant parties, submit to Congress recommendations regarding how to promote distance education through digital technologies while 'maintaining an appropriate balance between the rights of copyright owners and the needs of users'.
- States explicity that 'nothing in this section shall affect rights, remedies, limitations, or defences to copyright infringement, including fair use.'

## Is copyright dead?

With the advent of anonymous file copy programs, such as Freenet (**www.sourceforge.com/freenet**), some predict the death of copyright. File-sharing programs such as Gnutella and Napster are also impacting on copyright power although those organizations that have an identifiable central organization can (and have) been prosecuted.

Such programs confirm that the main difficulty with copyright is the detection and prosecution of the offender. If anything can be safely and anonymously copied, so that the original copier cannot be traced, even with a governmental level of code-breaking resources, what is the point of copyright?

“THE MAIN DIFFICULTY WITH COPYRIGHT IS THE DETECTION AND PROSECUTION OF THE OFFENDER”

While the development of peer-to-peer file copy programs raises interesting concerns, it is too early to write off copyright. It has been, for example, fairly easy to copy software for many years, and indeed the software industry estimates that half its potential income is lost to pirate copies, mostly through friends sharing copies of programs with each other – does this include you? However, copyright remains important in the industry, and by and large commercial organizations of any scale respect it.

The publishing, music and video industries are being affected in a host of dramatic ways by the internet, although copy protection, or lack of it, is one of the major issues. Digital rights management (DRM) is a hot topic, and there are many methods being developed to make copying harder, although there is always a battle between the defenders and the hackers, and between assurance and ease of use. In addition alternative methods of payment for content, such as the Street Performer Protocol **www.counterpane.com/street_performer.html** are being developed. The protocol is a method for getting payment before publication. It works like this. The author publishes a sample of the work, such as a track or a first chapter, together with a statement that if more is wanted contributions should be sent to some honest banker, together with a target amount and date for the next instalment. If the target amount is reached by the date, the next episode, chapter or track is then released, for example under the GPL. If not, the money so far contributed is returned. This does not remove the role of the publisher, but changes it so that promotion and revenue is collected before, rather than after publication.

## Trademarks

Registration of a trademark gives right to exclusive use of a name or mark in relation to the goods or services for which it is registered.

A trademark is any sign which can distinguish the goods and services of one trader from those of another, *and* be represented graphically. A sign includes words, devices (logos), three-dimensional shapes and sometimes sounds and smells. One of the earliest trademarks in the UK that still exists today is the red triangle for Bass's beers.

A trademark is therefore a 'badge' of trade origin. It is used as a marketing tool so that customers can recognize the product of a particular trader. Famous examples include 'Coca-Cola', Microsoft and the Nike 'swoosh'.

To be registrable a trademark must:

- be distinctive
- not be deceptive
- not conflict with other marks.

A trademark is issued for a particular class or classes of goods or services. For example, one of my companies once had the trademark 'Helios' for an operating system that we published. 'Helios' was also a trademark for sun-cream, and several other things. We were granted the trademark because people buying sun-cream were unlikely to confuse it with software.

When you apply to register a trademark you must specify the goods or services on which the mark is used, or is proposed to be used. Unlike a patent, there is no need for secrecy before applying. The goods or services must also be listed by reference to the class or classes of goods and services in which they fall, and in terms that can be clearly and easily understood without being misinterpreted.

Trademarks can also be used for certification marks, such as the BSI Kitemark, and for collective marks such as membership of some industry group.

Trademarks are registered and must be applied for in each local jurisdiction. However, you can now apply for a common European Community mark, and an application to WIPO (the World Intellectual Property Organization, **www.wipo.org** ) will enable you to register your trademark in countries who have signed up to the Madrid Convention, although that does not yet include the US. In the US, trademark registration is on a state-by-state basis, and prior use of a trademark, such as a sale invoice, must be shown in each state to enable you to register it. Generally once a trademark is registered, you must use it, or risk losing it.

‘TRADEMARKS ARE REGISTERED AND MUST BE APPLIED FOR IN EACH LOCAL JURISDICTION’

It is important to note that a company name does not imply a trademark, or vice versa. Similarly a trademark does not imply a domain name, nor a domain name a trademark, although there recently has been some case-law which has suggested that a domain name can be held as illegal ‘passing-off’ if it relates to a website under a similar or same name as a trademark, which offers similar goods to that trademarked.

One problem highlighted by the web is what might be a trademark for one company in one country, may refer to and be owned by a different company in another country.

## Domain names

Domain names are controlled by the Internet Corporation for Assigned Names and Numbers (ICANN), and the various registrars appointed by them. They were and are issued on a first-come, first-served basis. Ownership of a domain name does not imply or require a company name or a trademark, and with a few special exceptions, such as the .gov domain, they are open to anyone. This has led to people ('cybersquatters') attempting to register many domain names and then sell them at exorbitant prices to the slower-moving brand owners. Few have got rich, and the market is not what it was. The courts have taken a dim view of attempts to pass-off as the genuine article a website with the same domain name as a well-known brand, and equally have defended legitimate prior use by others that happen to have the same name or initials as a well-known brand.

## Other IPR

There are other formal rights, such as registered designs, which protect the visual appearance of an object, and design right, which protects the shape and configuration of an object, which tend to be used in specialized industries.

Design protection takes two forms. Registered designs give stronger protection but require application to the Patent Office, rather like a patent. Design right gives weaker but automatic protection, without the need for registration. Its automatic nature means that it has many similarities to copyright. Full details can be found on the Patent Office website.

## Non-disclosure agreements

Non-disclosure agreements (often called NDAs) allow you to discuss your bright new idea with people such as potential customers or investors, without running the risk of this being a publication for patent purposes or of them stealing your idea and exploiting it without you. An example non-disclosure agreement is given at the end of this chapter.

These agreements are usually mutual, and allow exchange of specific information for a particular purpose, although they can be worded quite generally. NDAs impose quite severe obligations on the parties involved to keep the matter confidential, and in some cases record a complete list of staff who have had access to the information. If possible, it is best to avoid NDAs that require any verbal information to be put in writing and explicitly recorded,

‘NDAS IMPOSE QUITE SEVERE OBLIGATIONS ON THE PARTIES INVOLVED TO KEEP THE MATTER CONFIDENTIAL’ or that require much record keeping, as these can impose a severe burden. NDAs should be worded so that they also cover any casual information obtained, for example from overheard conversations, or oversight of documents while the visitors are on your premises. Companies should put in place a suitable register of NDAs, and a way of monitoring their obligations, such as the duty to return or destroy documents no longer required.

Some companies refuse to sign NDAs on the grounds that they are more trouble than they are worth, and that they have a sufficient reputation for trustworthiness – either you trust them enough to do business with them or you don't. For a small company, an NDA is of doubtful value in any case, as such a company is unlikely to have enough resource to sue a major company for infringement with any chance of success. Large companies will often require you to sign their standard form of NDA, and will not accept any other format.

## Valuation

Valuation of IPR can be difficult, especially where the IPR is in the nature of a trade secret, rather than something more formal. Valuation can be based on a number of different criteria:

- what the IPR cost to develop
- what the IPR would cost the opposition to develop
- the value of the business potential
- the value of the physical assets.

Valuations can vary by several orders of magnitude, but still be 'correct' in some sense. It all depends on the circumstances, and whether one is buying or selling. While a valuation normally assumes a willing buyer, a willing seller, and a 'going-concern' business, the situation can be different. For example, half-finished software in a forced sale (fire sale) is worth virtually nothing, but the same IPR in a takeover might be worth many millions.

## IPR and employment

There is a presumption that IPR generated by an employee or director in the course of their employment is automatically transferred to the employer. This usually covers work done in company time or with company equipment, or using existing company information. However, some companies go further and require that any invention made during the course of employment (including the novel you are writing on you own home wordprocessor) is the property of the company. It is not clear however, whether this would be enforceable.

IPR is slippery stuff. All employees and directors have a duty of confidentiality, either written explicitly in their contracts or unspoken, although key staff with knowledge of the secret can still leave and even go and work for the opposition. Some employee and director contracts prohibit employment by competing companies, or by companies within a certain radius within a particular time period. How enforceable these prohibitions would be is open to debate. Clearly, when you leave a company you will carry away with you in your head, all of the additional experience, expertise and knowledge acquired while working for the company and this, of course, is allowed for. What is not allowed is anything that you could carry away in your briefcase, such as plans, customer lists, or trade secrets.

‘IPR IS SLIPPERY STUFF’

### Exercise

**one** Do an informal online patent search about your idea, using the resources below. If you don't yet have an idea you want to research, choose a topic, for example 'digital television'.

## Further reading

**www.delphion.com/** Was the IBM patent database search site

**www.patent.gov.uk/** UK Patent Office

**www.uspto.gov/** US Patent Office

**www.european-patent-office.org/** umm ...the European Patent Office

**www.wipo.int/** The World Intellectual Property Organization

**www.york.ac.uk/org/auril/aurilsec.htm** Association of University Research and Industry Links (AURIL)

**www.btgplc.com/** British Technology Group

**www.luna.co.uk/~patmg/** Patent and Trademark Group (professional body for people who search patents)

**www.businesslinks.co.uk/** Business Links (DTI)

**www.cordis.lu/ips-helpdesk/** IPR Helpdesk (EU Commission)

**les-europe.org/** Licensing Executives Society

## *Patents*

**www.cipa.org.uk/** The Chartered Institute of Patent Agents

**www.bl.uk/services/sris/pinmenu.html/** Patent information centres, British Library

**www.bl.uk/** British Library homepage

## *Registered trademarks and company names*

**www.itma.org.uk/** The Institute of Trademark Agents

**www.companies-house.gov.uk/** Companies House

**www.tradingstandards.net/** Trading Standards offices

## *Copyright*

**www.patent.gov.uk/dpolicy/index.html/** Copyright Directorate

**http://eblida.org/ecup/** European Copyright Users' Forum

**www.cla.co.uk/** Copyright Licensing Agency

**www.law-services.org.uk/** The Law Society

**www.alcs.co.uk/** Authors' Licensing and Collecting Society

**www.prs.co.uk/** Performing Rights Society

## *Registered designs*

**www.pro.gov.uk/** Public Record Office

**www.design-council.org.uk/** Design Council

## *Domain names*

**www.icann.org/** The Internet Corporation for Assigned Names and Numbers

**www.nominet.org.uk/** Registry for UK names

**www.internic.org.uk/** Internic (registrar and whois services)

## Example NDA

**Disclaimer** The sample document is for general information purposes only so that readers may gain an understanding of the common legal documents and issues that technology companies may encounter. These materials do not purport to provide full legal advice on the area concerned and the author and publisher do not accept liability for loss resulting from the use of or reliance on these materials. The sample document and other information in this section of the book are written from the perspective of English law. Other legal jurisdictions may treat these transactions differently.

If you intend to use any of this information or related documentation for a specific purpose then specific professional legal advice should be taken.

CONFIDENTIALITY AGREEMENT

**THIS AGREEMENT** is made on the day of

**BETWEEN**:

(1) [... ... ... ... ... ... ... ... ... ...] (the 'company')

**AND**

(2) [ ] of [ ] ('[name]')

each of the above being together referred to in this Agreement as the 'Parties' and individually as a 'Party'.

**NOW IT IS HEREBY AGREED** as follows:

**1. Introduction**

The purpose of this Agreement is to record the terms upon which:

(a) The company is prepared to disclose to [name] certain information relating to the company in connection with discussions relating to [... ... ... ..] (the 'Discussions'); and

(b) [name] is prepared to disclose to the company certain information relating to [name] and, if appropriate, its parent undertaking and any subsidiary undertakings (as defined in sections 258 and 259 of the Companies Act 1985) thereof in connection with such Discussions.

**2. Definition of Confidential Information**

For the purpose of this Agreement 'Confidential Information' means and includes:

(a) information of whatever nature relating to the company or to [name] (as the case may be) and their respective customers, businesses or financial affairs which has been obtained prior to the date of this Agreement and/or is obtained during the term of this Agreement either in writing or orally from or pursuant to the Discussions;

(b) those portions of analyses, studies, reports and other documents prepared by any Party to this Agreement which contain, evaluate or analyze any such information as is specified in paragraph (a) above; and

(c) information of a commercially sensitive nature relating to the company or to [name] (as the case may be) obtained by observation during visits to the Parties' premises.

**3. The Commitment**

In consideration of the Confidential Information being made available by each Party to the other, each Party hereby irrevocably undertakes to the other to use the other Party's Confidential Information and otherwise to act in accordance with the terms and conditions hereinafter contained. The disclosure of Confidential Information by one Party to the other shall in no way be construed to imply any kind of transfer of rights connected with the Confidential Information including, without limitation, any trademarks or business secrets. The obligations of each Party in respect of Confidential Information made available pursuant to this Agreement shall continue notwithstanding the termination of the Discussions.

## 4. Safekeeping

Each Party will treat and safeguard as private and confidential all of the other Party's Confidential Information and will take all reasonable precautions in dealing with any such Confidential Information so as to prevent any third party from having access to the Confidential Information.

## 5. Limited internal dissemination

Each Party will only disclose or reveal any of the other Party's Confidential Information disclosed to it to those of its personnel (which term shall for the purpose of this Agreement include employees, directors, officers, agents of and consultants to the relevant Party) and tax, legal and financial advisors who are required in the course of their duties to receive and consider the same for the purpose for which it is supplied. Prior to the disclosure of any Confidential Information to any such personnel or tax, legal or financial advisor such Party will inform them of the confidential nature of the material and of the provisions of this Agreement.

## 6. Non-disclosure to third parties

Save as otherwise permitted neither Party will at any time without the other Party's prior written consent:

(a) disclose the other Party's Confidential Information to any third party, either directly or indirectly; or

(b) disclose to any person either the fact that discussions or negotiations are taking place between the Parties or the content of any such discussions or negotiations or any of the terms, conditions or other facts with respect to the other Party, including the status thereof,

unless required to do so by law or by the order or ruling of a Court or tribunal, judicial or regulatory body or recognized stock exchange of competent jurisdiction, in which case, if the disclosing Party is required to disclose such information it will, unless prohibited from doing so, notify the other Party promptly in writing of that fact and in any event, prior to making such disclosure.

**7. No direct approach to personnel**

Neither Party will without the prior written consent of a director, officer or authorized representative of the other Party discuss the other Party's Confidential Information with any of the other Party's personnel save as may be necessary for the proper performance of the services to be provided by each Party to the other.

**8. Limitation on further actions**

8.1 The Parties agree that they will use the other Party's Confidential Information solely for the purposes described in the Introduction to this Agreement.

8.2 It is understood that all communications regarding the Discussions, requests for additional information or meetings or questions will be submitted or directed to authorized representatives of the relevant Party.

**9. Exclusion from Confidential Information**

These terms and conditions will not apply to any Confidential Information which:

(a) is in or becomes part of the public domain or is or otherwise becomes public knowledge by any means other than by breach by either Party of any obligation contained herein; or

(b) was previously or is at any time hereafter disclosed to a Party by any third party having the right to disclose the same provided that such source is not known to the receiving Party to be bound by a confidentiality agreement with, or other obligation to secrecy to, the other Party; or

(c) is released from the provisions of this Agreement by written consent given by a director or authorized representative of the Party to whom such Confidential Information relates; or

(d) was otherwise independently acquired or developed by the receiving party without violating its obligations hereunder.

### 10. Return of Confidential Information

All Confidential Information relating to either Party ('the first Party') (including all copies held by the other Party) will forthwith be returned to the first Party upon receipt by the other Party of a written notice to that effect from the first Party and the other Party will destroy those portions of all copies of any analyses, studies or other documents prepared by such Party for its use containing, evaluating or analyzing any Confidential Information relating to the first Party or such portions of documents containing any analyses, studies or other documents and use commercially reasonable efforts to expunge and destroy any such Confidential Information or such portions of documents containing any analyses, studies or other documents from any computer, word-processor or other device in its possession or custody or control containing such information. If, for reasons of storage space or office management, either party wishes to return or destroy Confidential Information received from the other party, it shall notify the other party of its intention. Unless the other party objects, the Confidential Information and all copies thereof, whether in printed or electronic form, shall be returned or destroyed in accordance with this paragraph 10.

### 11. No responsibility for information provided

Each Party understands and acknowledges that neither the other Party nor any of its personnel is making any representation or warranty, express or implied, as to the accuracy or completeness of the Confidential Information relating to such Party, and that neither the other Party nor any of its personnel will have any liability to any person resulting from any use of the Confidential Information.

### 12. Breach of Agreement

Each Party acknowledges and agrees that damages would not be an adequate remedy for any breach of this Agreement and that any affected

Party shall be entitled to the remedies of injunction, specific performance and other equitable relief for any threatened or actual breach of this Agreement.

**13. Governing Law**

These terms and conditions shall be governed by and construed in all respects in accordance with the internal, substantive laws of England and the Parties submit to the jurisdiction of the English Courts for all purposes relating to this Agreement.

**IN WITNESS** whereof this Agreement has been executed by the Parties on the day and year first above written

**SIGNED** by )

for and on behalf of )

**[the company ]** )

**SIGNED** by )

for and on behalf of )

**[name ]** )

# 4 Writing the business plan

‘Chance favours the prepared.’

(Louis Pasteur, 1822–95)

THIS CHAPTER IS ABOUT GETTING YOUR IDEAS straight, both for your own benefit and so that you can convince others. There are lots of other people you will need to convince as you go along:

- the people who work with you
- the people who will back you and pay for the development, whether they are your employers, or the bank manager and investors
- the people who will use your product.

These people may not have your vision, or your background knowledge. You will need to explain and persuade them, and more importantly describe what you need from them, be it money, or effort, or their expert contribution.

The business plan is where you write down and explain what you are going to do, when and why.

## Venture capital criteria

Before we get down to the nitty-gritty of the actual business plan, it will be constructive to look at the criteria which might be used by a typical venture capitalist to evaluate whether to invest in your business. Even though most companies do not raise their initial funding from a VC, these criteria are a useful yardstick. They have been developed over time by looking at many businesses, and by analyzing the reasons behind why some businesses are successful and others fail.

The VC will need to be able to easily extract the data to make their evaluation from the plan that you submit, and it is surprising just how many plans fall at this basic hurdle.

The criteria are, in order of importance:

- global, sustainable, under-served market need
- strong management team
- defensible technological advantage
- believable plans
- 60 per cent internal rate of return (IRR) and exit route.

We will examine each of these in more detail.

## Global, sustainable, under-served market need

This is the single most important criterion. Each word refers to a particular quality and all are important.

❝THERE IS NO MARKET FOR BETTER MOUSETRAPS IF MICE ARE EXTINCT❞

**Market need** The first thing that you need to establish is whether there is, or will be, a market for whatever goods or services you propose to supply. There is no market for better mousetraps if mice are extinct.

Typical questions you should ask are:

- Who needs it? Why?
- What is the profile of a typical customer?
- How many potential customers are there?
- How many of those will buy? At what price?
- How does this number change with price?
- What are your potential customers' top three key requirements? What are their concerns?
- How will these customers buy your product (route to market)?
- How will they find out about you?

## SWOT

SWOT stands for Strengths, Weaknesses, Opportunities and Threats. Factors such as competition, business opportunities or new markets are often analyzed using the SWOT methodology, which provides a framework for thinking about them.

### *Strengths*

S is for the *internal* strengths of the proposition. What are the good points? What is unique for this proposition? What do you do well? Maybe it is a particular talent, a strong team, or unique IPR? Don't be modest, but be realistic.

### *Weaknesses*

W is for *internal* weaknesses. What are the bad points? What do you do badly? What could be improved? Maybe you are short of resources, or lack marketing skills. Again be honest, but realistic.

### *Opportunities*

O is for *external* opportunities. How can you exploit your strengths? What are the trends in your favour? Maybe some external event has opened a window of opportunity, or there is a new market that you can exploit? Again, be realistic, but creative.

### *Threats*

T is for *external* threats. Which of your weaknesses could stop you performing? What are the obstacles in your path? The biggest threat is usually the competition, but other risk factors are market changes, technology change and timescales. And don't forget dear old Murphy – see http://dmawww.epfl.ch/roso.mosaic/dm/murphy.html.

We looked in Chapter 2 at market research, and some techniques that can help establish answers.

**Under-served market** How is the problem solved today? What are the existing solutions? Why are they inadequate? Does the market need to be created, and if so, why and how?

What and who are the competitors? Be sure to think about those who could enter the market in the future as well as those that are currently active in the marketplace. What are the strengths and weaknesses of the competition? How well resourced are they? How quickly can they adapt to change?

**Sustainable** The concept of sustainability was introduced originally by Bains & Company the management consultants. Is your product or service a one-shot wonder, or is it something that people will go on buying?

‘THE CLASSIC EXAMPLE OF A NON-SUSTAINABLE PRODUCT MIGHT BE THE EVERLASTING LIGHT BULB’

The classic example of a non-sustainable product might be the everlasting light bulb; if a company proposed to manufacture such a thing, it might expect good initial sales, but these would soon fall off as anyone needing a light bulb would have already purchased one, and the market would saturate. Markets for products such as TV sets and word processors are saturated with sales for such goods reduced to a small replacement level. With products like these, the manufacturers, and indeed the industry, must strive to innovate and constantly bring out new ‘must have’ features like high definition television (HDTV) to drive new sales, or else try and diversify to new markets.

**Global** With the increasing globalization of markets, your new product or service must have global potential, or at least be compatible with the global market. The internet, for instance, does not stop at national borders. You need to be aware of how your product will be perceived in different countries and in different languages.

## Strong management team

**You can't do it all by yourself** There is a very real limit to the size of project you can tackle all on your own. A public company must have at least two directors by law. Far from diluting your vision, your partners bring skills and knowledge you may not have. These are people you will need to recruit and motivate, and build into an effective team. We look at team building in Chapter 9.

A typical core team, using US terms (UK terms in brackets) consists of:

- Chief Executive Officer (UK: Managing Director)
  - — Responsible for the day-to-day running of the company.
  - — Responsible for formulating policy proposals and implementing the Board's decisions.
- Chief Financial Officer (UK: Finance Director)
  - — Keeps the books and is usually a qualified accountant.
  - — Prepares management reports and budgets.
  - — Advises on fundraising.
  - — May also act as Company Secretary – keeps Board minutes, official papers and shareholder records.
  - — Runs the company administration. Human Resources (Personnel) may report to the CFO or direct to the Board, as may Site Services, the section that keeps the physical infrastructure of the office running, managing things such as telephones, stationery and computers, and not forgetting cleaning and keeping the coffee machine full.
- Chief Technical Officer (UK: Technical Director)
  - — Does the technical work. In a large and established organization this role may split into Chief Operating Officer, responsible for manufacture, shipping and day-to-day operations, and Chief Technical Officer, responsible for new developments and technical strategy.

- Vice President of Marketing (UK: Marketing Director)
  - — Responsible for deciding what to sell, how much to sell it for, and how to reach the customer, involving advertising, distribution routes, and market targeting methods.
  - — Monitors feedback from the customer helpdesk and deal with other customer-facing aftercare.
- Vice President of Sales (UK: Sales Director)
  - — Shifts the product. Runs the sales force.
  - — This job may be combined with the Marketing Director role in a small organization, but the two jobs involve separate skills. See Chapter 13.

In a small organization jobs are often combined. The CEO may also be the CFO for financial prudence, or VP of Sales and Marketing for a customer-focused company. However, it is unusual for the CTO to also be the CEO, as these tend to be different skill sets.

**Alliances alliances** Very few non-trivial projects can stand on their own and most depend on relationships with others, such as a particular supplier, distributor or route to market. You may also have important relationships with other companies, and could rely on them for success, if for instance you manufacture components that others build into a finished product.

‘THE WHOLE IS GREATER THAN THE SUM OF THE PARTS’

These relationships need to be recognized and exposed. The whole is greater than the sum of the parts, and showing that the critical relationships are in place makes the plan considerably more real and believable.

**Professional advisers** One way to gain extra knowledge and experience is to make use of professional advisers. All companies are required by law to have professional accountants as auditors, and you will undoubtedly have a bank manager, so make use of these people that you already pay!

However, there are many other sources of advice, some of which are low cost or free, such as small business advisers provided by the government or your local community. Lawyers, accountants, merchant banks, market research companies and management consultants will all compete for your business and money. Such advisers can be extremely expensive, but if picked wisely, can be a useful asset, so choose and use with caution. Make sure the personal chemistry works, as they will be intimately involved. Also be sure to keep them fully and accurately informed – their advice can only be as good as the information they get.

**Open source** A special sort of collaboration is the open source movement (http://opensource.org), where many hackers collaborate informally in loose alliances based on common, freely available source code. This has been shown to be a particularly effective methodology for the development of some very large software products, although it is not a universal panacea. Whether this methodology can be used deliberately for a new development is still an open question.

What is more likely to happen is that, as the project matures, conventionally structured companies spring up to exploit the common heritage, for example by making particular versions of their own and offering consultancy based on the common tools. The classic example is Linux and companies such as Red Hat Software, which has recently come to market with a valuation of many billions of dollars and in turn sponsoring further open source development.

‘WHAT IS MORE LIKELY TO HAPPEN IS THAT, AS THE PROJECT MATURES, CONVENTIONALLY STRUCTURED COMPANIES SPRING UP TO EXPLOIT THE COMMON HERITAGE’

## Defensible technological advantage

The market should be *defensible*. You need to be able to defend your market share from attack by others who are seeking entry, and a share of your market. From a different perspective, what have you got that no one else has? Are there any barriers to market entry that you can overcome more easily than your potential competitors?

Defensible technological advantages may include:

- *Intellectual property*, for example patents, copyright in key programs (see Chapter 3)
- *Defensible technological leadership*. Even without formal protection, you may have trade secrets, established relationships and skilled staff, and be faster on your feet than other potential new entrants to the market and therefore able to out-innovate them.
- *Brand value.* You may be able to establish a brand that your customers trust and value, which acts as a barrier to entry by others. Establishing and maintaining a brand is expensive and time consuming.
- *Niche market*. If the market is sufficiently small and specialized, it may not be worthwhile for a competitor to enter in opposition. When a niche market is combined with a strong brand this may be enough to deter competitors completely. However, even here you cannot afford to go to sleep as there are always bright young people with aspirations, and companies in nearby niches eyeing your territory.

## Believable plans

If you have not completed the plans, set out how you intend to develop them. The VC will want to see that you have put some thought into them, and are not attempting the impossible. They will check that the resources you estimate are reasonable for the task in hand – often they are dramatically underestimated. A typical set of plans is given in the section about doing the writing later.

## Financials

A typical VC will look for about 60 per cent IRR (internal rate of return – see Chapter 5 for a definition). Put another way, a rough rule of thumb is that the business should plan to generate enough cash to be able to repay the original investment in the third or fourth year of operation.

Special considerations apply in industries with long asset life, such as cable communications and drug development, where both the development cycle and the cash generation is over a longer period. Here EBTDI (earnings before tax, depreciation and interest) is used as a short-term measure of growth.

Special considerations also apply at present to internet-based industries, where market valuation is based in part on traffic and the number of customers, in the belief that this can be turned to profit in the medium term. Time will tell if this is realistic.

## Exit routes

Your investors will want the assurance that there will, eventually, be some way of realizing their investment. However, starting a business with the sole intention of exiting is one sure way to fail, and it is impossible to predict right at the beginning of the business what its eventual fate will be. None the less a paragraph or so to show that you are aware of the possibilities and have concerns for the eventual outcome for investors, with some indication of timescale, will not come amiss.

"STARTING A BUSINESS WITH THE SOLE INTENTION OF EXITING IS ONE SURE WAY TO FAIL"

We discuss valuation methods in Chapter 17, and exit routes in Chapter 18.

# Doing the writing

We can now look at writing the actual business plan. Remember the business plan describes what you want to do, and is the primary document to help convince others to help you. Reflecting the discussion above, appropriate chapter headings might be:

"WHATEVER YOU CAN DO OR DREAM YOU CAN DO, BEGIN IT. BOLDNESS HAS GENIUS, POWER AND MAGIC IN IT. BEGIN IT NOW"

(JOHANN WOLFGANG VON GOETHE)

1 Executive summary and funding requirement
2 The concept
3 The market
    3.1 Global market size and need

3.2 Sustainability

3.3 Competition

3.4 Marketing plans

4 The team

4.1 CEO

4.2 CTO

4.3 CFO

4.4 VP Sales and Marketing

5 The technology and its IPR

6 Summary of plans

6.1 Development plans

6.1.1 Methodology

6.1.2 Milestones

6.2 Marketing

6.3 Sales and distribution

6.4 Quality and industry standards

7 Financials

*Appendices:*

1 Financial model

2 Key staff

3 Letters of support

4 Correspondence re IPR

5 Full development plan

6 Full marketing and sales plan

7 Examples and brochures

## *A note on style:* KISS: Keep It Simple Stupid!

Write for the target audience. The people reading the plan will include bankers and lawyers. They don't have the benefits of your insight and experience, or your deep knowledge of the subject. If your ideas and plans are worthwhile, they should be capable of simple explanation without patronizing the reader. You could even consider having different versions of the plan for different target audiences, although the difficulty of keeping them co-ordinated means that it's better to have a single, well-written document.

Say what you want clearly and directly in the executive summary, and what the expected returns might be. It is surprising how many business plans fail because it's too hard to figure out what they are asking for.

“SAY WHAT YOU WANT CLEARLY AND DIRECTLY IN THE EXECUTIVE SUMMARY”

The British Venture Capital Association Handbook (**www.bvca.co.uk**) is a source of good information about venture capital in the UK. Its *Guide to Venture Capital* (**www.bvca.co.uk/Publications/index.htm**) gives good advice on writing a business plan, and the directory gives the names and addresses of venture capitalists who might fund your vision. This and other sources of finance are discussed in Chapter 6, and non-disclosure agreements and legal frameworks are discussed in Chapter 7.

### The elevator pitch

You are alone in an elevator with a potential investor. You've got 30 seconds between floors to sell your idea. What do you say?

The discipline of honing your ideas down so you can deliver the message in 30 seconds is a good one, and helps you to concentrate on the essentials:

- What is it?
- Who's going to buy it?
- Why?
- How much money will you need?

# Business plan sections in detail

## Executive summary and funding requirement

The people who will read your plan are busy people. A VC will get hundreds of proposals to read. This is the section where you grab their attention, sell the idea, and stand out from the crowd. It is also the section that can be copied into internal summaries and other working documents.

You need to explain your idea in less than a page or so and state the main points of the plan, such as alliances or particular risk factors. Not least, you should state how much money you need or whatever else you are asking for, and what the payback might be. It is amazing how many plans omit this.

You need some housekeeping pages such as an index, change control, confidentiality statement, list of advisers (accountants, auditors, bankers, lawyers), and other quick reference data, such your name, address, phone number, e-mail and other contact details. Your lawyer may propose some rubric as well.

## The concept

This is the introduction to your plan, where you explain the big idea. You can also describe the history of the project so far, and just why now is a good time.

## The market

Who needs your product, and why? The section on criteria under 'Market need' above sets out the questions that need to be answered.

This is the most important section, and deserves care. Quote key external findings and reports, although bulk market research data, brochures and fine detail are better as an appendix. What are the barriers to market entry for yourselves, and for others?

## The team

Key players and their roles. Don't worry if you have not yet identified or recruited all the staff, but make sure that you show that you are aware of the need. One of the ways a VC can help is in finding the right people to work with you.

Also mention any alliances, other investors or existing or potential contracts, but put the actual letters of support in the appendix.

## The Technology and IPR

Describe the technology, distinguishing what already exists and what needs to be developed.

‘ONE OF THE WAYS A VC CAN HELP IS IN FINDING THE RIGHT PEOPLE TO WORK WITH YOU’

If it needs to be developed, how will it be done? Is it defensible: what is the IPR, who owns it, and what is to stop others developing it?

- **Summary of plans.** Detailed plans explain the nitty-gritty of how you intend to implement your ideas. As they are often long and complex, they should be in an appendix, bound separately, with only a short summary here. The plans should be inter-related and should include sections such as:
- **Development.** Shows how you intend to develop the product or service from where you are now. Details any technical development needed, together with setting up, manufacturing or other facilities.
- **Marketing Plan.** Illustrates how you intend to market your product or service, covering marketing activities, including market research and market communications such as advertisements, mail shots, websites and reporting structures. It should also include estimates of market size and segmentation
- **Sales Plans.** Details sales activities, including direct sales activity, and distribution methods. It should also include detailed sales estimates, with justifications for the numbers.

- **Quality and industry standards.** Explain how you will build quality into your product or service. What testing will you put in place? How will you check the documentation? What aftercare will you have? What happens to returns and customer complaints?

## Outline financials, growth and exit

Key assumptions and a one-page financial summary. What are your intentions for the business, and how will the investors get their money back?

This section will contain financial projections in the form of a predicted profit and loss statement, cash flow, and balance sheet, typically for a five-year period. The level of detail will depend on the particular plan, but there will be less detail the further into the future the prediction is made. For example, the first year will have monthly figures, the second and third year quarterly, and beyond that the fourth and fifth year will have annual estimates.

‘AN INDICATION OF POSSIBLE EXIT ROUTES CAN BE GIVEN’

An indication of possible exit routes can be given, but at this early stage these can be no more than expressions of intent. None the less, the emphasis should not be on the exit, but on buildng a viable business.

The bulk of the detail will be in a separate appendix, with a one-page summary in the main body of the report, and a few lines in the executive summary.

## Appendices

The appendices may be bound as a separate volume. They will include the following:

- marketing support and collateral (brochures, surveys, etc.)
- CVs of key staff
- IPR, patents, etc.
- letters of support or contract intent
- plans in full
- financial projections.

## Exercises

**one** Choose a ground-breaking product with which you have some familiarity, for example development of the ARM processor, or the introduction of the Apple Mac, or the development of a drug. Write a short outline business plan that the pioneers might have written at the start of the project, but disguise the name. Swap with a colleague and critique each other's plan. Would you have funded their plan, without the benefit of hindsight?

## Further reading

Blackwell, E., (1998), *How To Prepare a Business Plan : Planning for Successful Start-up and Expansion*. Kogan Page

British Venture Capital Association (BVCA) *Directory of Members*. **www.bvca.org.uk**

British Venture Capital Association (BVCA) *A Guide to Venture Capital*. **www.bvca.org.uk**

Nokes, S. (2000) *Startup.com*. FT/Prentice Hall

# Money and legal affairs

# 5 Creating the budget

> ‘Money makes the world go around
> The world go around
> The world go around
> Money makes the world go around
> It makes the world go ’round.’
>
> (*Cabaret*, lyrics by Fred Ebb, 1966)

THIS CHAPTER IS ABOUT CASH. SOME MAY REGARD ALL money as theft, and love of money as the root of all evil. Although the kingdom of heaven may run on love, companies on earth run on cash. There are severe limits as to what can be achieved in a purely barter economy. In this chapter we discuss raising, controlling and eventually sharing out that cash.

The key lesson is that cash, not profit, is king. Cash flow is far more important than short-term profitability, especially for start-ups and internet companies. Running out of money and not being able to pay the bills will surely bring down even the most promising company, yet there are some apparently successful high-tech companies that have enriched their founders while profit remains a distant dream. Never mind the profit, at least in the short term, but don't run out of cash.

‘THE KEY LESSON IS THAT CASH, NOT PROFIT, IS KING’

This chapter is a brief introduction to the some of the principles and technical terms associated with accountancy. It is not intended to teach you everything about accounting. You will not learn to be an accountant – for that you should read specialist books, take appropriate courses and serve your apprenticeship. However, this chapter may allow you to have a more intelligent conversation with your accountant or specialist adviser. Subsequent chapters discuss raising money and stocks and shares.

# Introduction to accounting

## Why have accounts?

Why bother keeping accounts at all? After all, some businesses seem to get along just fine without. For example a small dealer keeps a large wad of cash in his back pocket. When he buys he pays cash and when he sells he takes cash only. The thickness of the wad tells him how he is doing. However while this might work in a single-person cash-only business, real enterprises, to say nothing of the taxman, need more sophisticated measurements.

### Accounts are instruments on the dashboard of the company

Accounts tell you (and others) how you are doing. They do not control the company, but hopefully they provide reliable measurements. They act as a scoreboard of your performance. They are an important part of the communication chain, both internally within the company, and externally to interested people such as:

- shareholders and investors, both existing and potential
- employees and employee organizations such as unions
- government and especially the Inland Revenue
- business analysts
- your competition
- the public
- other parties, such as creditors and charities.

**To control, you must first measure and record** Accounts provide a way of meaningfully comparing performance over time and between different activities. They also provide a means of comparison between actual performance and predicted budget.

**Statutory duty** The law says you must keep accounts, and as a director of a company, you could, in principle, go to jail if your accounts are not kept, or are inaccurate.

## Legal requirements

The legal requirements of running a business are as follows.

*Keep proper books of account*

Books must be kept in a recognized format and be a true, honest and fair statement of the company's account. You must publish, every year:

- the trading (profit and loss) account
- the balance sheet
- a cash-flow statement (historic).

*Annual audit*

You must have your books audited annually by a recognized and licensed practitioner, to provide independent confirmation of how well or badly you are doing.

*Solvency*

You must be able to pay your bills. Even if you do not have the money in the bank, you must have a justifiable belief that you can either raise the money or make sufficient profits to enable you to pay your bills as they fall due.

If you cannot pay, and have no prospects of being able to do so, you have a statutory duty to declare bankruptcy and shut the company down. Failure to do so is fraud, for which you can go to jail.

## Avoiding error: double entry

In days of yore, before computers, people kept accounts by hand. Double entry bookkeeping was devised as a way of checking against error. The principle is simple; since money doesn't disappear, everything is entered twice, once as the source and once as the sink. If done correctly the whole thing should add up to zero at the end of the period. Think of it as a network made up of accounts. Money (or value) moves from one account to another, but does not magically appear or disappear – at least not in properly managed companies. The whole system is self checking.

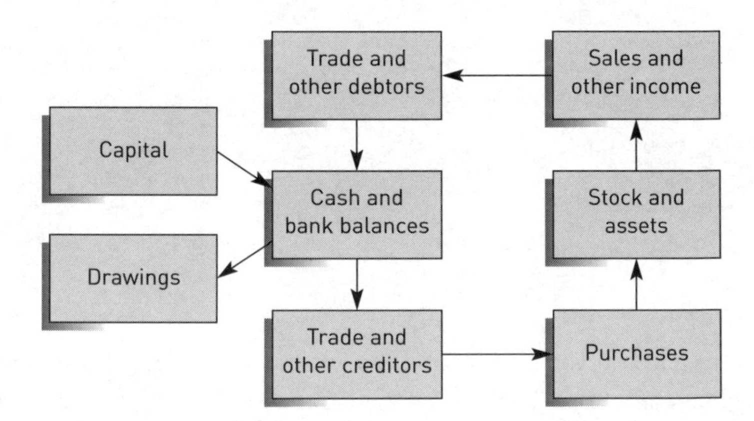


When something is bought, money is taken out of the bank account, reducing the value in this account but increasing the value of the stock held. If something is sold, the value of the stock is reduced, but the money in the account is increased.

In the old days accounts were kept in big ledgers. On the left-hand side were positive entries – *debits* – that is money received or due to be received. On the right hand side are *credits*, that is money which has been paid or is owed. The two totals should be equal, if need be by a balancing figure representing the transfer to another page.

You need to know about accounts so you can do the financials in your business plan. Once you start on the business, you will need accounts to see how you are doing against your plan and to comply with the requirements of your finance providers and the law of the land.

Account name Ledger page number:

| Date | Description | Amount | | Date | Description | Amount |
|---|---|---|---|---|---|---|
| | Transfer in<br><br>Debit side | | | | Credit side<br><br>Transfer out | |
| | Total | | | | Total | |
| | | | | | | |

However, unless you want to make accountancy your career, all you need follows. This is the 'five-minute MBA' accounts course. We will explain the simple arithmetic to accounting, some little bits and pieces you won't find anywhere else and then the three spreadsheets which your business plan requires. The more advanced bits and the legal compliance bits can be left to a proper accountant.

## Simple accountancy

Rule: Do not question the way accountants do things. Years of experience have shown it works. Learn it parrot fashion. And one day you will find it comes automatically

Accountants have just two vertical views of the world, which they refer to as ***debits*** and ***credits***. Now repeat: 'debits and credits'. Do not say 'credits and debits' as that is wrong. There are also two horizontal views: every debit and every credit will ultimately appear in either a ***balance sheet*** or a ***profit & loss account***.

It would be helpful if you would learn this little table:

## Profit & loss account

| Debit | Credit |
|---|---|
| *Cost of goods sold* (all goods for resale minus any stock left at the time) | *Sales* (invoices raised or, in a retail scenario, cash received) |
| *Expenses* (all the costs including wages) | |
| ***Profit*** (always a balancing figure) | |

## Balance sheet

| Debit | Credit |
|---|---|
| *Fixed assets* (e.g. computer, car) | *Creditors* (people you owe money) |
| *Debtors* (people who owe you money) | *Loans* (banks you owe money) |
| *Stock* (goods for resale) | *Capital* (the money you put in) |
| *Bank* (assuming a positive balance) | *Retained profit* (the profit made so far) |

Finally, learn a simple little rule: If you increase a debit then you must *either* increase a credit *or* decrease another debit by the same amount. Similarly, increasing a credit either requires an increased debit or a reduction to another credit. This rule ensures that all the debits in total equal all the credits in total, always.

Congratulations! You now know about accounting.

So let's do a worked example:

**1** Open a bank account with £1000 to start your business.

Debit: Bank £1000

Credit: Capital £1000

**2** Go to market and write a £600 cheque for some mushrooms.

Debit: Stock £600

Credit: Bank £600 [We could say Debit: Bank -£600 but instead we copy what real accountants do with minus numbers and change debit to credit]

Quick check on the bank – We put £1000 in and spent £600, leaving £400.

In accounting speak debit £1000 then credit £600 leaves a debit of £400

**3** Door to door we sell half the mushrooms for £700 which we pay into the bank.

Debit: Cost of goods sold £300 (half of £600)

Credit: Stock £300 (reducing stock for what we sold)

Debit: Bank £700

Credit: Sales £700

We can then do some accounts:

## Profit & loss account

| | | | |
|---|---|---|---|
| Cost of goods sold | £300 | Sales | £700 |
| Profit (=balance) | £400 | | |
| | £700 | | £700 |

## Balance sheet

| | | | |
|---|---|---|---|
| Stock | £300 | Capital | £1000 |
| Bank | £1100 | Retained profit | £400 |
| | £1400 | | £1400 |

**4** The mushrooms are looking old – we sell the remainder to a caterer for £350.

Debit: Cost of goods sold £300 (being the rest of the stock)

Credit: Stock £300

Debit: Bank £350

Credit: Sales £350

Now our accounts look like this:

## Profit & loss account

| | | | |
|---|---|---|---|
| Cost of goods sold | £600 | Sales | £1050 |
| Profit (= balance) | £450 | | |

## Balance sheet

| | | | |
|---|---|---|---|
| Stock | £0 | Capital | £1000 |
| Bank | £1450 | Retained Profit | £450 |

If you have followed so far, you will have seen that debits seem to add and credits seem to subtract. If doubly observant, you will have noticed that both 'Capital' and 'Sales' have credit totals but are shown as positives.

Accountants will tell you that it is shown as a positive because it is a positive credit shown on the credit side of the page, but we prefer to consider it positive more for show as a lot of minus signs or brackets would be messy.

With all this addition and subtraction and with debits equalling credits it is no wonder the computer has revolutionized accounting.

The money put into the system represents the capital of the business, made up from the shareholders' investment, loans made to the company, and accumulated profit, less any drawings (payments to the shareholders) made.

The balance sheet is where all the threads are brought together. As its name implies, it should balance – the money coming in must match the money going out. It shows the source and the application of funds.

The money flowing round the business is the working capital, and is represented by stock, and trade debtors, that is people who owe the company money for goods or services purchased. It is the money needed for day-to-day operations.

‘THE MONEY FLOWING ROUND THE BUSINESS IS THE WORKING CAPITAL’

Some of the capital will have been spent purchasing fixed assets, such as plant and machinery, or computers and office furniture. The value of these decreases only slowly, and this is reflected in the accounts. A depreciation charge, representing, say, one-third of the value of the computer is made from the capital account to the profit and loss account each year. Thus the value of the asset, in our example a computer, is decreased to zero over three years. Different companies have different policies on how long things last, but three years for computers, and five years for more solid stuff, like office furniture, is usual. It does not matter so much what the rules are, as long as they are clearly stated in the accounts.

## Principles of accounting

**Boundaries** In order for your accounts to be of any use they must have well-defined and consistent boundaries of what is being accounted for. Without consistency the accounts will be meaningless, and it will be hard to track progress, or to compare with other companies' accounts. Important points to define clearly are:

- The *entity*, which is, for example, the company, or department to which the accounts refer. The entity needs to be the same in each period so that the accounts can be meaningfully compared. If the boundaries are not the same each time, you are counting sand.
- The *periodicity*, the period to which these accounts refer to, such as a month or a year. Accounts are a snap shot of activity over a period of time, and there must be strict cut-off dates, both at the beginning and end of the period. Comparison is made much harder if the accounts refer to different lengths of time.
- The *basis of valuation*, that is, how things are valued. For example, a piece of software might be valuable when finished, but of no value if incomplete. A computer might cost a certain amount to buy, but be of little value in a forced sale. The normal assumption is that the entity will continue as a *going concern*, that is the entity will continue to trade in the future, and that future sales are not forced. Thus it might be reasonable to depreciate the value of a computer over its useful life, say three years, and to value a piece of partly completed work-in-progress software as the value of the effort so far expended.

**Measurement** Measurement must be *quantitative*, that is, everything must be reduced to a common numeric basis of comparison, usually monetary value. Intangibles, like goodwill, should not be included, unless you have concrete evidence of what someone would pay for them and that a deal will follow.

**Ethics** Accounts should be 'true and fair'. There is little point in deceiving yourself, and deceiving others is fraud. If in doubt, understate profits and overstate losses. It is far better to be prudent, and have pleasant surprises, than to be over-optimistic and have an unpleasant time adjusting to reality. Accounts should be objective and supported by external evidence, rather than personal estimates.

### Accountancy programs

Of course, nowadays, an organization of any size will keep electronic rather than manual accounts, with the system semi-automatically set up and the accountancy program providing the smarts. None the less, it all relies on the principles just outlined.

## Financial projections

### Forecasts and budgets

Forecasts are predictions about the future, and are used as a basis for planning. Budgets are financial plans for specific time periods, based on the forecasts for that period.

Forecasts can cover both the local issues, such as how many of each product the company might sell, or how many people it will employ, but can also cover larger issues such as the state of the economy, or the price of energy. Large-scale forecasts are sometimes based on PEST analysis, standing for Political, Economic, Social and Technological issues.

Budgets are very important, and help a company to plan, control and implement decisions. According to an old saying 'if you fail to plan, you plan to fail'. Now that the majority of everyday bookkeeping is automated, budgeting is probably the accounting activity that you will be most involved in.

A budget is a stake in the ground. It may be a wild guess, but at least it is some sort of estimate, which you can improve as you get more information or experience. Write down your assumptions, and then you can see how well they were justified. Clearly the closer to the present, the finer grain and more accurately you can budget.

Budgets are often done as a *rolling budget*, for example predicting the next year by monthly results. Every three months the results of the last quarter are compared against the budget, and the budget rolled forward by three months so it continues to predict the following 12 months.

A budget is the closest thing to a crystal ball that you will get. A budget is not a one-time-only thing, but is a living, breathing entity at the core of your business.

Adopt an attitude of 'pessimistic realism'. You don't want to be so pessimistic that nothing is worth doing, but you should make conservative estimates, so that the surprises are nice ones. Tell the truth – know the worst.

**Comparison with actual** A budget is not a static thing, but, like a garden, needs constant tending.

Budgets are useless if left to moulder. They are most valuable when you compare them with the actual results at regular intervals, usually monthly or quarterly. There are two important questions that you should ask yourself:

- Why are the results different? They probably will be. Is this because of some unforeseen event?
- Why was the budget wrong? Maybe some of your assumptions were incorrect, or the model was wrong? Only by comparison with actual will you get better at prediction. At first your budgets are likely to be wildly out, but don't be discouraged. Persevere, and, as you gain experience, they will become more accurate. Successive refinement is much easier than original thought. I've learnt over the years that I'm always too optimistic and that I should double the costs and halve the income from my first guess.

Since the advent of tools like spreadsheets, the mechanics of budgeting has become much easier. The reader is urged to be familiar with the use of spreadsheets such as Microsoft Excel, now pretty well the defacto standard.

**Sensitivity analysis** Sensitivity analysis is a technique used to assess how sensitive the budget plan is to changes in the assumptions on which it is based. Any particular budget is one snapshot. More importantly it is a model of your company, and if you flex some of the parameters you can gain more insight. You can begin to see how sensitive your profit is to the various parameters. For example, in a typical restaurant, food costs are only about 30 per cent of expenditure, so profitability is not strongly sensitive to raw ingredient costs, compared to say, percentage occupancy.

What happens if you only sell half as much as you expect? Or twice as much? What happens if raw materials cost more, or the project over-runs? Suppose you put the price up by ten per cent: you might not sell as many meals, but because you will need less labor and overheads, you may make more profit.

In these examples:

- The boundary is the project.
- The periodicity is one month.
- The basis of valuation is that no value is assigned to work in progress
- Depreciation of capital equipment is, for the moment, ignored, as are interest payments on loans and overdraft.
- Overheads are assumed as 100 per cent of salary.

## Profit and loss

A profit and loss account tracks how much money you have made or lost and is therefore one of the most important financial accounting statements. For reasons lost in the mists of time, accountants usually write negative numbers, like losses, in brackets.

Here is an example of drawing up a profit and loss budget statement.

## Example

Suppose you are fortunate enough to have won a software contract worth £100,000. Payment is on fairly generous terms – 30 per cent on signature, 30 per cent halfway and 30 per cent on delivery, with 10 per cent retained for six months against bugs. The software is expected to take two people six months to complete. For simplicity, let's presume that programmers cost £30,000 per annum, and that everything else (heat, light, rent, stationery, postage, communications costs, hire of computers, fees, insurance) is the same as salary, that is overheads of 100 per cent, which is about normal for the industry.

We can draw up a budget as follows:

**Example profit and loss budget**

| Month | 1 | 2 | 3 | 4 | 5 | 6 | 7 | 12 | Total |
|---|---|---|---|---|---|---|---|---|---|
| Income | 30,000 | | 30,000 | | | 30,000 | | 10,000 | 100,000 |
| Expenditure | | | | | | | | | |
| Programmers | 5,000 | 5,000 | 5,000 | 5,000 | 5,000 | 5,000 | | | 30,000 |
| Overheads | 5,000 | 5,000 | 5,000 | 5,000 | 5,000 | 5,000 | | | 30,000 |
| Total costs | 10,000 | 10,000 | 10,000 | 10,000 | 10,000 | 10,000 | 0 | 0 | 60,000 |
| Profit in the month | 20,000 | (10,000) | 20,000 | (10,000) | (10,000) | 20,000 | 0 | 10,000 | 40,000 |
| Profit to date | 20,000 | 10,000 | 30,000 | 20,000 | 10,000 | 30,000 | 30,000 | 40,000 | 40,000 |

So far so good. Looks nice and profitable, and we should take the contract.

## Cash-flow statement

The above statement represents *a profit and loss* budget, and presumes that income is accounted for when it is invoiced, and that payments are accounted for as soon as the invoice is received. Unfortunately the world is not quite that simple. We are not living in a cash economy, and many companies take up to 60 days to pay, that is two months after the month in which the invoice is issued, which can be how long it takes to work through their accounting system. However your programmers still want paying on time, otherwise they walk. Assume you will pay other invoices after 30 days.

The picture now looks like this:

## Example cash-flow budget

| Month | 1 | 2 | 3 | 4 | 5 | 6 | 7 | 8 | 12 | Total |
|---|---|---|---|---|---|---|---|---|---|---|
| Income | | | 30,000 | | 30,000 | | | 30,000 | 10,000 | 100,000 |
| Expenditure | | | | | | | | | | |
| Programmers | 5,000 | 5,000 | 5,000 | 5,000 | 5,000 | 5,000 | 30,000 | | | |
| Overheads | 5,000 | 5,000 | 5,000 | 5,000 | 5,000 | 5,000 | 30,000 | | | |
| Total costs | 5,000 | 10,000 | 10,000 | 10,000 | 10,000 | 10,000 | 5,000 | 0 | 0 | 60,000 |
| Cash flow | (5,000) | (10,000) | 20,000 | (10,000) | 20,000 | (10,000) | (5,000) | 30,000 | 10,000 | 40,000 |
| Cash in bank | (5,000) | (15,000) | 5,000 | (5,000) | 15,000 | 5,000 | 0 | 30,000 | 40,000 | 40,000 |

The totals are the same, but now we need to go to the bank manager and arrange a £15,000 overdraft for *working capital*.

Worse still, we have not allowed for contingencies. Suppose the project takes two months longer than we anticipated in getting to the half way stage. Now the cash-flow budget looks like this:

## Example cash-flow budget

| Month | 1 | 2 | 3 | 4 | 5 | 6 | 7 | 8 | 9 | 10 | 16 | Total |
|---|---|---|---|---|---|---|---|---|---|---|---|---|
| Income | | | 30,000 | | | | 30,000 | | | 30,000 | 10,000 | 100,000 |
| Expenditure | | | | | | | | | | | | |
| Programmers | 5,000 | 5,000 | 5,000 | 5,000 | 5,000 | 5,000 | 5,000 | 5,000 | | | | 40,000 |
| Overheads | | 5,000 | 5,000 | 5,000 | 5,000 | 5,000 | 5,000 | 5,000 | 5,000 | | | 40,000 |
| Total costs | 5,000 | 10,000 | 10,000 | 10,000 | 10,000 | 10,000 | 10,000 | 10,000 | 5,000 | – | – | 80,000 |
| Cash flow | (5,000) | (10,000) | 20,000 | (10,000) | (10,000) | (10,000) | 20,000 | 10,000) | (5,000) | 30,000 | 10,000 | 20,000 |
| Cash in bank | (5,000) | (15,000) | 5,000 | (5,000) | (15,000) | (25,000) | (5,000) | (15,000) | (20,000) | 10,000 | 20,000 | 20,000 |

As mentioned, normal accounting convention puts negative numbers in brackets. In this scenario, we needed to arrange a £25,000 overdraft, and won't get half the profit until eight months after we have finished work.

For both small and large enterprises, cash is much more important than profit. Running out of money and not being able to pay the bills will surely bring down even the most promising company. This is not to say that profit is unimportant, but if you run out of cash you won't be around to reap the rewards.

### Balance sheet

We can now draw up a projected balance sheet, for example at Month 9 of the last example.

## Example balance sheet as at beginning of Month 9

| | | |
|---|---|---|
| *Fixed assets* | | |
| Computers | 10,000 | |
| Furniture | 3,000 | |
| *Current assets* | | |
| Work in progress | 10,000 | Retainer, not yet invoiced |
| Trade debtors | 30,000 | Amount invoiced, but not yet paid |
| Cash | 0 | Normally there would be some petty cash |
| *Less: Current liabilities* | | |
| Trade creditors | 5,000 | |
| Bank overdraft | 15,000 | |
| *Net current assets* | 20,000 | |
| *Total assets* | 33,000 | |
| *Representing* | | |
| Proprietors' capital | 13,000 | |
| Plus: Accumulated profit | 20,000 | |

The £13,000 of proprietors' capital represents the cash put up initially to buy the computers and the furniture. Normally there would be an element of depreciation reducing their value, and hence the overall profit, as well as interest payments on the bank overdraft, but these have been omitted for simplicity.

## Fixed and variable costs

It is useful to distinguish in the accounts between fixed and variable costs. Fixed costs are things like the rent on the factory or office that stay more or less fixed except in the extremes, regardless of how many units you make. Making one more unit won't change the cost. Variable costs are things like the cost of materials that vary directly with the number of units that you make.

## Taxation

Corporation tax would become due on the annual profit. There are complex rules about when corporation tax needs to be paid, and what relief can be counted against it that your accountant can give you advise about.

If the company was registered for VAT, VAT would need to be added to the invoices, and subtracted from goods purchased, and the difference in VAT paid to the Excise. You need to register for VAT if the company turns over more than £52,000, but the Chancellor of the Exchequer changes the rules and the limit every year. The VAT man will come and inspect your accounts, and has draconian powers to fine the company if everything is not in order, or late.

For internet trading VAT can be difficult to calculate correctly, as the amount you need to charge will depend on the address, and possibly the nationality of the purchaser. Different arrangements are needed if the customer is in the UK, or the EU, or the US or the rest of the world. Your accountant or friendly VAT inspector may be able to help.

PAYE and National Insurance need to be deducted from the wages and paid to the Inland Revenue promptly.

Do not neglect to pay the tax and VAT due. Render unto Caesar that which is Caesar's. More companies have been made bankrupt by the unforgiving taxman than by any other cause.

## How much do you need?

How much you will need depends on your particular business, and there is no substitute for constructing a budget on the lines above. Unfortunately it is becoming more and more expensive to start a small start-up. This is in part due to the cost of developing increasingly sophisticated new products, but also the growing costs for aspects such as required consumer safety tests and marketing.

As a rule of thumb, you probably need to figure on not having any external revenue for about a year. As a rough guide to how much you will need to start up a business:

- For a single-person consultancy, working mostly from home, you might need somewhere in the order of £100,000 to cover living, travel, marketing and things like a computer and part-time secretary, book keeper and some professional advice from lawyers and accountants.
- For five people, offices, some plant or machinery, stock, marketing, etc. you might need funding in the order of £1 million.
- For 20 people, starting a small factory, developing and marketing a mass-market product might need £5 million capital.

## Tests of financial standing

Given a set of accounts, we can apply a number of tests to see how we are doing.

Typical tests are given in the table overleaf.

## EBITDA (Earnings Before Interest, Tax, Depreciation and Amortization)

For companies that have a large investment in long-term assets, such as infrastucture companies with a long market development cycle, measures such as EBITDA which exclude capital may give a more accurate measure of the company's performance.

| *Name* | *Value* | *Result* | *Comment* |
|---|---|---|---|
| Current ratio | Current assets / Current liabilites | 2 is good, 1.5 acceptable but < 1 indicates potential cash-flow problems. | Measures liquidity, that is how easily a firm can pay its short-term debts. |
| Acid test (Liquid ratio) or Quick health check | Liquid assets (Current assets – Stocks) / Current liabilities | 1 is acceptable but < 1 indicates potential cash-flow problems. | This is a more severe than the current ratio: assumes that you may not be able to sell stock in emergency |
| Gearing | Net borrowings (long-term debt) / Shareholders' funds (equity) | A gearing ratio of 50 per cent is equal to a debt/equity ration of 1:1. This ratio shows your reliance on borrowings, that is, what percent of your finance comes from debt. A low gearing is desirable. | Shows vulnerability to interest rate rises, and affects how much money the shareholders make. Highly geared companies are vulnerable in bad times, as the interest on the debt must be continued to be paid. |
| Return on investment (ROI) | Profit before tax / Shareholders' funds | Efficiency – 40 per cent for sustainable high growth. | Profitability test. |
| P/E (price/ earnings ratio) | Share price/net profit per share | High-growth companies should have P/Es between 10 and 20 in normal times. | Expectation of future profitability. |

## Auditors, accountants and other advisers

You are required by law to keep your accounts in order, and therefore it is advisable to seek help with your accounting. On an annual basis you must appoint professional auditors to certify your accounts. But many small firms employ a bookkeeper, perhaps on a part-time basis, to keep their books in order. As you grow, the bookkeeper, now full time, may double as office manager, but eventually you will need a full-time, properly qualified chief financial officer, complete with a full accountancy staff.

Even if you are a qualified accountant, you will probably want to appoint a firm of accountants to advise, and to help you set up your accounting systems. They can also help you draw up budgets, and generally keep you on the right path. These accountants will need to be independent from your auditors. If you are raising money, the assistance of a reputable firm of external accountants is a great help in lending credibility to your estimates and your enterprise.

‘EVEN IF YOU ARE A QUALIFIED ACCOUNTANT, YOU WILL PROBABLY WANT TO APPOINT A FIRM OF ACCOUNTANTS TO ADVISE, AND TO HELP YOU SET UP YOUR ACCOUNTING SYSTEMS’

Accountancy is dominated by the large firms (the ‘big five’). Don’t be afraid to approach them, for example by getting in touch with their local office. Although their fees are high for a new start-up, they are realistic and realize that new and small companies can’t afford that much. What they really want is to be able to charge you large fees when you have become successful. If they like your project they may well come to some arrangement to reduce or defer their fees, or even take an investment in the company. Some run or have relationships with incubators specifically to help start-ups. If you don’t ask, you won’t get.

Your advisers such as accountants and lawyers are your friends, not your enemy. They are on your side, and want you to succeed. They have seen the inside of more companies than you have, and know what works and what does not. Tell them everything, and listen to their advice. They are not just there for show.

## Interpreting accounts

The ability to accurately understand a set of accounts is essential, for example to decide whether to invest in a company.

Here is a semi-structured approach to interpreting company accounts:

*1. Quick look:*

Take a quick look at the profit and loss statement, balance sheet, and cash flow. You are trying to get an overall picture at this stage, looking at things like company size, profitability, cash generation, etc.

*2. Financial summary trend*

Look at the financial summary in the accounts. Is the company growing, stagnating or declining? (See BCG categorization below.)

*3. Notes to accounts*

Read the notes to the accounts. What do these say about sales trends, different parts of the business, geographical spread, and any other special factors. Is there anything unusual that might mask some deeper problem?

*4. Chairman's statement, reports, etc.*

These are usually simply platitudes, but they can give a handle on recent events, contingent liabilites, and so on.

*5. Auditors' report*

Big warning bells should ring if the auditors' report is at all different from the usual unqualified approval. However it is normal and acceptable in smaller companies for the auditor to base some of their evaluation on the statements of the directors. This might apply, for example, to the value of work in progress.

*6. Profit and loss, balance sheet, cash flow*

Now look at the details: trends, ratios, etc.

Compare the results to industry norms are there marked differences, and if so why are they different?

*7. Write down your conclusions*

## Moving towards profitability

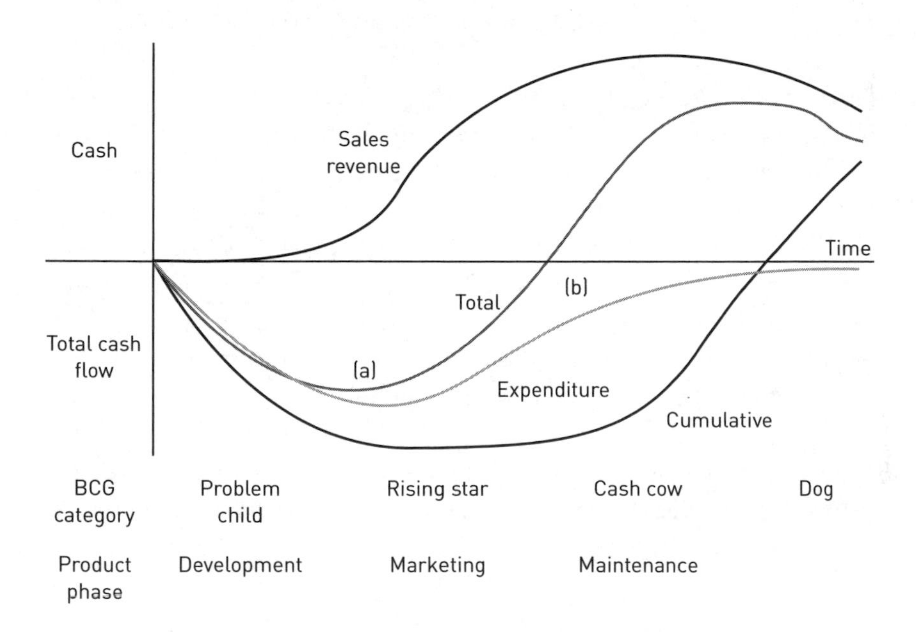


- The diagram shows cash flow for a typical product over time.
- The sales revenue curve is the actual income.
- The expenditure curve is how much is spent each period.
- The total curve shows the net income in the period – the difference between revenue and expenditure.
- The cumulative curve is the cumulative sum of the total curve, showing overall profit or loss.

At the beginning of the project, money is spent developing and getting the product to market, and also on the initial marketing. After a while, sales start to pick up, development costs decline, and the project becomes cash positive day to day (point (a)). That is, more money comes in than is spent in further development, market development and cost of goods. However, the project has still not made money over all, as there is a considerable accumulated back log of costs to repay. If sales continue in this fashion, the project will move into overall profit at point (b). Eventually sales will decline, or support costs rise to an unacceptable level, and the product reaches the end of its life.

How long this life cycle takes can vary wildly with the product, but all agree it is getting shorter. For instance it took 20 years for 50 per cent of households in the US to get colour TVs but it took five years for them to get VCRs, and only six months to get Sony Playstations. Computer hardware changes quickest, Moore's well-known law predicting a factor of change of two every 18 months. System software changes a little more slowly: Microsoft introduce a new operating system roughly every two years. Application software generally lasts about five years, although some, like games, have lifetimes measured in only months.

However, data changes slowly, and companies need to keep records for seven years at least, and may have the same customers or employees for 30 years.

The Boston Consulting Group characterized products as:

- *Problem child*. The product has high potential, but needs development and hence investment. Originally this was a product with little market share, but in a fast-growing market, but it can also be applied to early-stage development. Hopefully it will turn round, not just be a dog.
- *Rising star*. A product with demonstrated potential, but needs further investment and nurture. Originally a product with high share in a rapidly growing market.
- *Cash cow*. A mature product in its profitable phase, but not one that will last forever. Originally a high market share of a stagnant market.
- *Dog*. A product going nowhere. Originally a low share of a poor market. Can be a product with declining sales at the end of its life.

To see these plotted as a matrix go to
**www.bcg.com/this_is_bcg/mssion/ growth_share_matrix.asp**.

## The value of money

Money has a value that varies with time. A pound in your pocket today is worth more than the promise of a pound in a year's time.

*Example*

I have good and bad news. The bad news is that your elderly relative has died. The good news is that they have left you £100,000. However there is more bad news because you have found out that due to legal procedures it will take a year to get the money, although you are sure to get it. The good news is that you have a sympathetic bank manager.

You want to buy a fancy car now, borrowing the money against your inheritance. How expensive a car can you afford? Suppose the bank will lend you money at an annual interest rate of 10 per cent. In a year's time you will have to use the £100,000 to pay back the loan plus interest. If we write x for the amount you can borrow, then:

$$100{,}000 = x + 0.1*x, \text{ so } x = 100{,}000/1.1 \text{ or about £}90{,}909.$$

So £100K in a year's time is actually worth only just over £90K today. This is its *present value*, or *PV*.

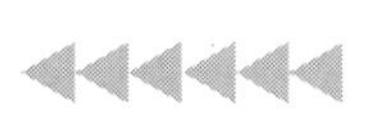

In general

$PV = I/(1 + r)^n$, where

I is the amount
r is the interest rate for the periods
n is the number of periods.

This notion, that the value of money depends on when you get it, is fantastically useful. It's what keeps banks and the whole of the financial community afloat.

Now it turns out that your relative had an insurance policy, of which you are the beneficiary, with another £100K coming your way. However, it will take two years for you to receive this money. How much can you afford now?

Here we have two years' interest to pay, so the PV is:

$$£100{,}000/(1.1)\times(1.1) \text{ or about } £82{,}640.$$

To find out how much we can afford in total, we can just add up the present values to give a *net present value* or *NPV* of £173,549. We can add them up because they have been referred to the same time value, so we are adding like to like.

Now suppose someone wants to sell you an investment that will pay you £100K in a years time, and another £100K a year later; how much should you pay? It's the same NPV problem: £173K – you can add up NPVs, as they have been referred back to the same baseline.

To take a larger example, suppose you strike oil in your back garden. Geologists can tell you how much oil there is and how fast you can pump it out. Economists believe they can tell you what price it might fetch in the future, or rather the likely range of prices, and hence the income potential over the years for maybe the 20-year lifetime of the well. Engineers can tell you how much it will cost to build the well, the pipelines, the operating costs, and not forgetting the final decommissioning costs. You can gather a set of positive and negative numbers representing income and expenditure for each year of the life of the project. To find out if the project is worth doing, (or maybe how much to bid for the mineral rights) all of these must be reduced to their present value and added up to give a net present value.

Of course, in a real example you would want to include risk factors, and allowances for possible changes in interest rates. These financial models can become very sophisticated, and the people who understand them and can construct and interpret them are among the highest paid in the city.

The best known of these models is the Black-Scholes model for option pricing, published by Fischer Black and Myron Scholes in 1973, for which they subsequently won the Nobel Prize.

## IRR

The counterpart to NPV is IRR, the internal rate of return. This is the interest rate needed to generate a particular set of cash-flow numbers. Essentially the IRR can be calculated by driving the equation backwards to find an apparent interest rate, given an investment today and a set of income figures.

In other words, the IRR is the discount rate that gives an NPV of zero.

Suppose we have a project that needs an initial investment, and in due course generates periodic profits (and losses). The IRR is the effective return on investment. In our example above we have a cash flow of:

| Period | 0 | 1 | 2 |
|---|---|---|---|
| Amount | (173) | 100 | 100 |

(Remember, negative numbers are written in brackets by accounting convention). The example corresponds to an IRR of 10 per cent.

High-growth projects should expect an IRR of 60 per cent or so.

Here is an example of about 60 per cent IRR over five years:

| Period | 0 | 1 | 2 | 3 | 4 |
|---|---|---|---|---|---|
| Amount | (173) | 0 | 125 | 250 | 400 |

Most spreadsheets have a handy IRR function built in.

## Exercises

**one** From University of Cambridge Computer Science Tripos 1995 Paper 9 Question 9

Explain the difference between a profit and loss account and a cash-flow statement. Under what circumstances would they show the same figures?

A small software company is offered a development contract, valued at £100,000 (excluding VAT), with 10 per cent to be paid at the start of the contract, 30 per cent invoiced at the first milestone (estimated after three months), 50 per cent invoiced on completion, with 10 per cent to be retained for three months after completion, as a guarantee against errors.

The company estimates that the project will require six months' work from each of two staff, whose annual salary costs are £36,000 and £24,000 respectively. Other overheads are approximately 120 per cent of salary costs.

Draw up an outline monthly profit and loss account and cash-flow statement for the project, ignoring VAT and bank interest. Salary and overheads are charged to the project only while the programmers are actually working on it.

What is the eventual profit the company expects to make, if it undertakes the project, and how much working capital will the project require?

The effort in the project turns out to be underestimated, and the company delivers the first milestone one month late, and completes two months late compared with the original schedule, requiring both programmers to work for the extra two months. How has this affected the profitability and working capital requirement?

**two** Find, on the web, the last published accounts for a large company like Apple. Calculate the ratios given in the chapter. Discuss your conclusions.

## Further reading

Dyson J.R. *Accounting for Non-Accounting Students*. Pitman

Parker, R.H. (1982) *Understanding Company Financial Statements*. 2nd edn. Penguin

Ramsden, P. (1999) *Finance for Non-Financial Managers*. Hodder and Stoughton

# 6 Raising the money

‘Business? That’s very simple – it’s other people’s money’

(Alexandre Dumas fils, 1824–95, *La Question d’argent*)

THIS CHAPTER LOOKS AT PLACES TO RAISE THE MONEY. It is much better to gamble with other people's money if you can, although of course you will have to pay them for it.

First we look at the two different sorts of finance.

## Debt and equity

You can raise money for a project or a new company in two different ways, either as debt finance or as equity.

Debt finance is a loan, such as a bank loan. You hire the money. You borrow the money and promise to pay it back some time in the future, together with the interest. If it is a large amount of money, you may have to provide additional security, such as handing over the deeds to your house or other valuables to be held in escrow to guarantee the repayment. However, the amount you pay back to the lender is the same whether you succeed or fail. Even if your company is widely successful, they get no more than the original loan and interest.

Equity finance is very different. This involves selling a portion of your company, usually as shares. If the company is successful, your shareholders benefit, for example via future dividends, which are a share of the profit, and via an increase in the capital value of the company, which will increase the value of the shares. There is no promise to pay the money back to the shareholders, or to pay interest. If your company fails you can walk away. Equity finance is discussed more fully below.

The debt in debt finance may be represented in the form of a *bond*, basically an IOU from the company. The bond may have *coupons* attached – in the old days these were physical coupons that you could tear off and send in for regular interest payments. Bonds may be tradable, with a price that fluctuates in the market. This is because economic conditions such as bank interest rates vary, as does the perception of creditworthiness or risk of the underlying organization failing to honour its commitments. As a result of these variables, the price (and hence *yield*) of the bond will also vary, and so a market may be made. *Junk bonds* are bonds with very high yields (interest rates) for risky ventures. They were popular in the US a few years back for raising money for high-risk companies.

‘THE DEBT IN DEBT FINANCE MAY BE REPRESENTED IN THE FORM OF A *BOND*’

A *debenture* is a bond issued without an underlying security. It stands in line for repayment behind other bondholders and creditors. To sweeten the pill it may confer other rights, and in particular the right to convert to shares in the company. Thus, for example, the debenture holders may have the right to take over the company under some circumstances, for example if the company gets into trouble and the debenture is likely not to be honoured.

*Gearing* is the ratio of debt to equity finance. A company that is highly geared, that is they have borrowed a lot of money, will be vulnerable to

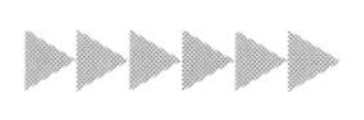

Interest rates are often measured as APR or annual percentage rate, although they may be quoted over a shorter interval, such as monthly, to make them seem less usurious. How the figures are compounded can make a big difference: an interest rate compounded daily is about double the same rate compounded annually.

Long-term debt is often available from your bank at 1–2 per cent over bank rate, say around 6– 7 per cent APR.

Credit cards, by contrast can charge rates in excess of 20 per cent APR.

Always read the fine print, and if in doubt get professional advice.

changes of interest rate. High gearing is bad in difficult times, as the company may be forced to use its money and profit to service the debt, that is pay the interest as it falls due rather than for growing the business.

## Shares

When you buy a share, you are buying a share of three different things. They normally, but do not necessarily, go together. They are:

- A share of control of the company. Shares may include voting rights that govern the ultimate control of the company.
- A share of the profits of the company usually represented as a dividend.
- A share of ownership of the assets of the company. This will have a bearing on the proportion of the company assets you are entitled to if the company breaks up. In particular, this will define the order of payment if the company ceases to trade and the assets are liquidated, assuming that there is surplus to be distributed.

A company may issue a number of different classes of shares, each with a different set of rights associated with them. A common class besides ordinary shares is that of *preference shares*, which stand in line before the ordinary shareholders for any distribution (but after the creditors and certain official bodies like the taxman). The equity counterpart of debentures are *redeemable preference shares*, which are preference shares that the company can repurchase at a fixed price if it is doing especially well.

The conduct and control of the company is governed by the Memorandum and Articles (Mem and Arts). These can in principle, say whatever the founders wish, although if they are too eccentric the company may have difficulty finding funding, bankers or advisers. Normally in the UK they are based on and refer to Table A of the Companies Act 1948 amended 1976.

Table A in the Companies Act sets out the entitlements of possession of certain critical percentages of voting shares for control (for normal Table A companies):

- Possession of 25 per cent or more blocks 'substantive' resolutions, such as a resolution to change the nature of the company or to wind it up.
- Possession of 50 per cent or more gives day-to-day control, including power to appoint and remove directors, but not to pass substantive motions.
- Possession of 75 per cent or more gives total control.

Which means if you hold less than 25 per cent of a private company, you are just along for the ride and cannot influence what the company does in any way. The directors and majority shareholders don't have any duty to report to you as a minority shareholder, except through the annual meeting and public accounts. None the less, the directors and majority shareholders do have a duty to not deliberately disadvantage the minority shareholders, but this protection is relatively weak.

‘IF YOU HOLD LESS THAN 25 PER CENT OF A PRIVATE COMPANY, YOU ARE JUST ALONG FOR THE RIDE’

Shares, and especially minority holdings in private companies, are only worth what a buyer is prepared to pay for them. While one day they may be worth a great deal, for example when the company as a whole is sold, or is publicly traded, for most companies until that point the shares are effectively worthless, and might just as well be used for wallpaper, as there is no market for them. Most private companies prohibit the sale or transfer of their shares without the agreement of the board, except in special circumstances, like death of the holder. Even then there may be pre-emption rights, forcing sale to existing shareholders, for example.

Therefore, be very wary if someone offers you less than a 25 per cent stake in a private company. It's a high-risk gamble, and one in which you have no influence on the outcome. Roughly 40 per cent of companies fail in their first three years, and only about one in ten turn into valuable enterprises. Even casinos offer better odds than this!

For public and listed companies there are other trigger points. For example, under the Stock Exchange rules anyone acquiring more than three per cent of a company's shares must declare whether they intend to mount a takeover bid.

## Sources of finance

Where should you go to raise the wind?

This will vary on how much you need to raise, and what you are prepared to give away. The following sources are listed in the order you might approach them, and the amount that you can raise from them. When you are just starting out you might raise modest amounts from family and friends, say enough to support you while you work out the business plan and do some initial market research. As the business gets more defined, you might attract angel finance, a bank loan and eventually one or two rounds of venture capital funding, before raising serious amounts of money from the City and the stock market.

### Family and friends

Family and friends are always the first port of call, especially in the very first stages. You need them on your side, anyway. Unless you have a rich relative, the amounts you can raise are relatively modest, but they might just keep you alive while you write the first draft of the business plan, and raise the next round of finance. Besides, who better to share your eventual success with, by letting them in on your enterprise at an early and low-cost stage.

The great advantage of family and friends, is that they are backing *you*, someone they know, and hopefully trust. They therefore do not need a great deal of convincing, and you can go to them at a very early stage when your big idea is still just a gleam in your eye. If they trust you, they may not need any security, and may be happy with relatively modest amounts of equity.

None the less, you will want to stay friends whatever happens. Be scrupulously honest, and make sure they know and understand exactly what they are getting into, and that they can afford to lose their investment if it doesn't work out – you don't want to take money that is vital to them. Make sure that the investment is documented, so that there is no argument later. Surprisingly, the arguments are worse and people's recollections differ more when you succeed and there is real money available to share, than if you fail. Get it written down while you are all still friends.

A version of family and friends are the various charitable and quasi-governmental schemes that now exist, such as (in the UK) the Prince's Trust and various local enterprise schemes. Your local job centre will be able to advise you on such schemes. It is cheaper and more politically acceptable for the government to give you a grant and to help set you up in business, than pay unemployment. The Prince's Youth Trust gives grants, soft loans (meaning loans with very relaxed conditions, and little expectation of repayment) and business advice to people under 25. They have a remarkable record of success.

'NONE THE LESS, YOU WILL WANT TO STAY FRIENDS WHATEVER HAPPENS'

## Angels

Angels are wealthy individuals, perhaps entrepreneurs who have sold out from a previous venture, who are prepared to invest in new start-ups. Angels bring not only cash, but also a wealth of experience and contacts that can be invaluable to your company. Finding them can be difficult, but your bank or other professional adviser may be able introduce you to some, or else keep an eye on your local newspaper for stories about local magnates, and then write to them.

Various groups, such as Digital People and First Tuesday hold regular meetings to introduce the wannabes to the haves. A note of caution: beware of anyone that attempts to take search fees up front for the introduction.

In the UK, and in some other countries, there are enterprise initiative schemes that make minority investment in new companies very tax-efficient, but you need to do the right paperwork. Your accountant and the local tax office can help with this.

## Banks

Your bank is the next port of call. They are a good source of debt finance for amounts from a few thousand to a few hundred thousand. Banks are willing to take a surprising amount of risk with new businesses. However they are in the numbers game, and more new businesses go right, for a while, than wrong. They benefit directly from your success, and from the extra money that you and your employees bring to the community. They also have access to cunning schemes, including:

- *Loan guarantee schemes*. These schemes are essentially insurance against default, but they allow the bank to lend greater amounts, or ask for less security than otherwise would be the case.
- *Factoring*. The bank can loan money against your receivables (the amount your customers owe, but have not yet paid), and in some cases against stock and work in progress. This can help dramatically with working capital requirements. Alternatively, you can sell your debtors (money owed to you) at a discount to the bank, which will collect the debts on your behalf, paying you the discounted amount immediately, thus freeing up your cash flow. This route could have financed the cash-flow example in the previous chapter.
- *Government enterprise schemes*. There are a number of schemes to encourage start-up businesses, which the bank can introduce you to.

In addition, the bank is plugged into the local business community, and your bank manager (sorry, they are now called relationship managers) may well be able to help with introductions. They will know the best lawyers and accountants in town for start-ups, and may be able to help in other and surprising ways. Larger banks have venture capital arms for equity finance, but that is later in the story.

Make no mistake; these people are not a charity. They are hard-nosed capitalists, and will want their pound of flesh. Interest rates will reflect the risk they are taking. They may also need additional security, and could require a mortgage on the deeds of your house, or a debenture on the assets of your company. They will want to ensure, as the saying goes, that you have enough skin in the game for it to hurt if you walk away. Hurt, but not cripple you, as that is not good for business, and they are unlikely in practice to actually sell your house from under you.

Keep your bank manager informed. They are in it with you, and are in a way a partner in the business. The one thing bank managers hate are surprises. Even if the news is bad, tell them. With enough notice they may be able to find ways round the problem, arrange extra finance, or extend the period of the loan. At short notice the range of options is very limited. The more notice and early warnings you can give, the better.

Note that a bank *overdraft* is really only designed to be a short-term loan, and a financial flywheel. The rates and conditions are not attractive for long-term borrowing, and if you need long-term money you would do better to negotiate a loan. There are many apocryphal stories about people financing start-ups from their credit cards. Although this may sound an easy option, it is definitely not a good idea, as the interest rates are crippling, and the loan terms short, much shorter than needed for the normal business cycle. Such stories merely indicate that the foolish people had not the wit to go and talk to their bank and get proper finance.

‘THE ONE THING BANK MANAGERS HATE ARE SURPRISES’

## Venture capitalists (VCs)

Despite their name, most venture capitalists are not particularly adventurous.

Venture capitalists exist to provide equity capital. Although there are a few funds that specialize in early stage investments (see the BVC *Directory of Members*, on their website), the majority of VCs will look to invest several million at a second stage or into things like the management buyout of an existing business, where there is an existing market and cash flow that can be quantified. A VC will have a three-to-five year horizon and within that time they will expect the company to be either acquired, or floated (listed on a public exchange), so that they can exit and get their money out. To this end, a large amount of capital can be made available, and the company grown rapidly. A VC's aim is to make considerable capital gains. However, the rapid growth needed may not be sustainable, and may not even be in the company's best long-term interests.

We looked at the criteria a typical VC might use in Chapter 4. Even so, a typical VC portfolio might have one or two stars, one or two complete failures, and eight or so 'walking wounded', which are companies holding their own, but not going anywhere.

A good VC provides more than just money. They have a range of contacts and access to expertise that can provide invaluable help to a fledgling venture. They have seen the inside of more companies than most. A good VC will ensure that the right people are in place for the rapid growth and subsequent flotation or sale, even if this means putting in their own experienced CEO, financial controller and possibly marketing and sales director over the head of the founder, if that is what it takes to make a fast-track successful company. The founder's role may change, but the entity and the investors benefit.

## Valuation

We discuss valuation at various stages in Chapter 17. It is useful to understand which are the events and activities that contribute most to value since, as a business owner, you will want to make sure that these activities are as successful as possible. The timing is also critical for bringing in investors. You want to bring in new investors just after you have achieved success in important value-enhancing activities. The investors will then be paying the 'new, higher valuation price' resulting from the advances that you have made with the business.

This approach also shows you why you don't want to raise too much money early in your new business's life. At early stages the business will be worth much less than it will be in the future; you will therefore have to exchange more equity for a certain amount of money at this stage than you will at a later stage when the business will be worth a lot more.

> 'THIS APPROACH ALSO SHOWS YOU WHY YOU DON'T WANT TO RAISE TOO MUCH MONEY IN YOUR NEW BUSINESS'S LIFE'

This is why there are usually multiple funding rounds for new businesses. You only want to raise as much money as you need to get you past the next major value-raising activity, at which point the business will be worth a lot more and you will get a much better price for your equity.

Of course, you must temper this with realism; you don't want to be spending all your time fund raising for small amounts. Typical funding rounds, timings and the amounts raised would be:

1 Initial work funded by business creators – typically £5–10,000.
2 Next stage to proof of principle, first business plan stage funded by friends and family – maybe up to £25,000.
3 Third stage may be seed-funding to produce a working prototype – typically £50,000.
4 First-round Venture Capital (VC) funding will take you to first customer orders–typical levels are £500,000–2,000,000.
5 Subsequent VC funding rounds may be needed for full-scale development and/or manufacture – the amounts in these rounds may be quite large.

Businesses that need more investment and capital, such as those in biotech or manufacturing, are likely to have more funding rounds than businesses with smaller investment needs, such as software firms.

## Raising money on the stock market

Companies can raise money by selling shares to the general public, usually via admittance to a listing on a recognized stock market. As we will see in Chapter 18, there are all sorts of complications associated with selling shares to the public, mainly to protect against fraud. It is a long and expensive process and, with a few exceptions, such as biotech companies, you need at least three years' trading record in order to sell shares to the public. It also complicates life: publicly listed companies need considerably more financial and reporting discipline than private companies. For example, any news that might affect the share price, such as the award of a large contract, or key staff changes, must be reported to the market first. In effect, the company now has two separate markets, one being whatever the business of the company is, and the other being its shares. This second market takes just as much attention as the main business of the company, demanding, for exam-

‘PUBLICLY LISTED COMPANIES NEED CONSIDERABLY MORE FINANCIAL AND REPORTING DISCIPLINE THAN PRIVATE COMPANIES’

ple, regular briefing of brokers and other city types, which is a considerable drain on senior time and resource.

Stock markets around the world have attempted to develop markets with lower entry requirements, such as AIM in the UK, NASDAQ in the US, the Troisième Marché in France, and the Neuemarket in Germany. In practice the entry barriers are still steep, and not to be contemplated unless it is worth raising the order of £10 million or so.

Floating the company on the stock market is used not only to raise capital, but also to give value to employee option schemes, and to provide an exit route, after a suitable lock-in time, for the original investors.

Despite a few well-publicized exceptions, such as Wit Capital, floating a company is not something you can do yourself. You will need lots of professional help, starting with a broker or merchant bank experienced in such transactions.

## Exercise

**one** Make a contribution table: list possible sources of finance for your idea, how much you could expect to raise from each, with what probability of success, and what the next action needs to be. For example:

| *Source* | *Amount* | *Probability* | *Action* | *Date* |
|---|---|---|---|---|
| Aunt Agatha | £10 | High | Go to tea | Next Saturday |
| Rich Uncle Fred | £1,000 | Medium | Write | Any time |
| Bank | £10,000 loan | Medium | Make appointment | Wednesday |
| Win lottery | £1million | Low | Buy ticket | By Saturday |

# Further reading

**www.digitalpeople.org** Digital People

**www.firsttuesday.co.uk** First Tuesday

**www.bvca.co.uk** British Venture Capital Association

Cary, L. (1998) *Raising Capital for the Smaller Business*. Seed Capital

Klein, A. D. (1998) *Wallstreet.com*. Henry Holt and Company (By the founder of Wit Capital.)

# 7 Legal issues

> "The Law is the true embodiment
> Of everything that's excellent
> It has no kind of fault or flaw,
> And I, my Lords, embody the Law."
>
> (W.S. Gilbert, *Iolanthe*, 1882)

THIS CHAPTER IS ABOUT THE VARIOUS OBLIGATIONS and duties you assume when you set up a company. It is based on current UK law, unless otherwise noted, although most jurisdictions have similar regulations. It is an introduction, not an authoritative text. If you are in any doubt you should consult your legal adviser.

## Perils of litigation

Stay out of court if at all possible. Lawyers are unbelievably expensive, and legal action is a great distraction and waste of your precious time, which is much better spent running the business. If you've got to the stage of court action, you've lost anyway, and whatever relationship you may have had has broken down. Taking someone to court is not going to fix whatever is bust, or get the contract completed, and probably won't even get them to pay. At best, legal action is a lottery and a high-stakes gamble, with no certainty of result. Whatever it is, it will take a long time, possibly years, by which time the world will have moved on, and only the lawyers will have got rich.

> ‘ONLY THE LAWYERS WILL HAVE GOT RICH’

That is not to say that should one suffer *any* wrong in silence without seeking a remedy, or one capitulates in the face of *any* claim, however unjustified, but that litigation is a desperate last resort, and not to be lightly undertaken. Principles are principles, but the costs of lawyers, not to mention your own time and effort, should discourage anyone from ever going to court. Settle out of court if humanly possible.

## Contract and tort

Lawyers find it useful to distinguish between contract and tort.

Contract covers voluntarily making the contracts you want to make, and avoiding the commitments you don't want to accept. We discuss contracts, and in particular some of the difficulties the internet introduces, below.

Tort is avoiding infringement of the rights of others, and giving adequate notice to others of any of your rights that you may want to enforce.

Examples, particularly relevant for e-commerce and websites include:

- *Defamation*. Defamation means publishing derogatory statements that you cannot prove true. Inclusion on a website, and in some cases just linking to others' statements counts as publication. A particular problem is with facilities such as bulletin boards, where others can enter statements, but because they are on your website, you are construed as the publisher. The terms and conditions of use need to make clear that you as the owner of the site accept no responsibility for the content.
- *Negligence* includes giving careless advice that might cause injury or (sometimes) loss. If you do not intend to give advice, make sure that the website says clearly (like this book) that the information is for illustrative purposes only, and should not be relied on for accuracy. Again, this is important for bulletin boards and similar websites.
- *Infringement of IPR*. Some include infringement of the rights of, for example copyright, trademark or patent holders as a tort.

## Regulations

There are many regulations with which you need to comply. Complying with regulations:

- avoids penalties
- ensures your rights are enforceable.

For e-commerce these include:

- Distance Selling: Consumer Protection (Distance Selling) Regulations 2000 (**www.hmso.gov.uk/si/si2000/20002334.htm**). Gives detailed rules on content of 'selling' web pages.
- Data Protection Act 1998 (**www.legislation.hmso.gov.uk/acts/acts1998/19980029.htm** or the Information Commissioner: **www.dataprotection.gov.uk/**). Requires people and companies that keep personal information to register. It gives rights of access and challenge, and limits the use or passing on of personal data without permission.
- Consumer Credit Act 1974. Lays out formalities for credit agreements, and financial protections for cardholders.
- Trade Descriptions Act. Requires accurate descriptions of goods and services offered for trade. In particular strong criteria are laid down for comparisons such as 'Half-Price Sale'.

## Duties of directors

The office of director in the UK carries with it stringent common law and statutory duties and responsibilities. There are roughly 250 or so separate laws that apply, including, but not limited to:

- Companies Acts
- Financial Services Act
- Shops Offices and Premises Act
- Health and Safety regulations
- Environment Protection Act 1990, and the Environment Act 1995
- Consumer Credit Act 1974
- Town and Country Planning Act 1990
- Fire Precautions Act 1971
- Discrimination Acts
- Data Protection Act

- taxes: VAT, ACT
- EU regulations: Distance Selling Directive
- etc., etc., etc. …

Failure to comply with these laws could result in heavy fines, disqualification from being a director of any company, or even imprisonment. Any director who thinks their company may have transgressed should take legal advice immediately.

Pretty well anyone can be a director, but certain people are excluded. These include:

“PRETTY WELL ANYONE CAN BE A DIRECTOR”

- bankrupts
- people with a court order banning them from being a director
- the company's auditors.

Directors are appointed and removed in accordance with the company's articles of association, but usually a director can resign at any time by giving notice in writing, although certain liabilities may remain for up to a year, and criminal liabilities can remain indefinitely. A director remains liable for his actions while in office, but can be removed by a simple majority of the votes cast by the shareholders.

A director can be, but is not necessarily, an employee, and it is useful to distinguish between the duties assumed by being a director from those of an employee. Directors can be, but are not required to be, paid separate fees.

Non-executive directors are not employees, and have no executive role within the company. Typically, they are senior figures that add their wisdom and experience to the Board of the company. They also have a role to play as independent observers, for example as part of the remuneration committee, deciding key salaries, or on the audit committee.

The basic function of the directors is to manage the affairs and activities of the company. In order to do this they must have certain powers. It is common in legal parlance to say that wherever powers go, obligations must follow.

The Board of Directors can do anything the company can, and can delegate authority to an individual. The actions of the company are limited by the 'Aims and Objects' clause of the articles of association. The position or status of a director is not a professional position, although a director is expected to exercise skill and diligence to a level that could reasonably be expected from a position with such responsibility. Of course, a director can make errors of judgement just like anyone else, but provided that the errors of judgement are reasonable, the director will not necessarily be answerable. However, if for instance, the company suffers loss as a result of its money being used by a director for purposes that it has not sanctioned, the director is liable to replace the money, however honestly the director may have acted.

Any employee, and in particular a director, has a fiduciary duty to the company. Broadly speaking, this means that a director must not knowingly do anything that will harm the interests of the company (which are not always the same as the interests of its shareholders). Directors have a duty of loyalty and good faith. There are three basic criteria that a director must satisfy:

‘DIRECTORS HAVE A DUTY OF LOYALTY AND GOOD FAITH’

- A director must act in the best interests of the company.
- A director must act as an ordinarily prudent person would act.
- A director must act only after reasonable enquiry, although they are entitled to rely on information given to them by officers of the company.

Other specific obligations that directors have as a result of their fiduciary duties include the following:

- A director must only use the powers which the company has conferred upon them. They must not exceed these powers, even if they think it is in the best interests of the company to do so.
- A director must not put themselves in a position where there is potential conflict between the interests of the company and their own interests.

> They cannot, for example, identify a business opportunity that would be valuable to the company, and instead use that information for their own benefit, or that of a competitor, such as setting up their own new company without offering the opportunity to the existing company. If you are a director, and think you may have a conflict of interests, get advice, and if setting up a new business get the permission of the company you intend to leave, or of which you are a director.

Directors are obliged to separate their own interests from those of the company. Where a director personally profits from their position, they must disclose it to the company.

Details of the duties of a director can be found in the excellent *Guide for Directors* from the Institute of Directors, or in one of the legal websites noted in the Further Reading section of this chapter.

## Employment law

This book is not intended to be a definitive guide to employment law. Employment law is very complex, and it differs from jurisdiction to jurisdiction. In the US, it even differs from one state to another. This chapter offers some pointers based on UK law, but if in doubt about your rights or obligations, consult a local professional adviser.

In addition to the topics touched on here, employment law covers such things as:

- collective bargaining and trade union law
- health and safety at work
- compensation in case of accident or injury
- maternity, paternity and sick leave
- minimum wage and pension rights
- taxation.

## Hiring and firing

**Employment contract/statement** Every employee must be given, within two months of starting work (one month if required to work abroad) a written statement of employment. This is not a contract, but may be taken to define the terms of the employment contract if there is no other documentation, and if the parties agree. It must include (taken from the DTI site: **www.dti.gov.uk/er/individual/statement-pl700.htm**):

- the *names* of the employer and the employee
- the *date* when the employment (and the period of continuous employment) began – these need not be the same, for example if the employee transfers from one company to another member of the same group
- *remuneration* and the intervals at which it is to be paid
- *hours* of work
- *holiday* entitlement
- entitlement to *sick leave*, including any entitlement to sick pay
- *pensions* and pension schemes
- the entitlement of employer and employee to *notice* of termination
- job *title* or a brief job description
- where it is not permanent, the *period for which the employment is expected to continue* or, if it is for a fixed term, *the date when it is to end*
- either the *place of work* or, if the employee is required or allowed to work in more than one location, an indication of this and of the employer's address
- details of the existence of any relevant *collective agreements* which directly affect the terms and conditions of the employee's employment – including, where the employer is not a party, the persons by whom they were made.
- if relevant, the *period* for which the employment abroad is to last
- the *currency* in which the employee is to be paid

- any *additional pay or benefits*
- terms relating to the employee's *return to the UK*.

**Employment from abroad** Employing a non-national will involve satisfying immigration requirements. Usually this means showing that there is no suitably qualified national available to do the work, by, for example, advertising the job locally and receiving no suitable national applicants – not just applicants who are not so good, but who could not do the work at all. There is much law on this subject, and the UK official guidance on work permits is given on **www.workpermits.gov.uk/**. In the EU, a national of any country belonging to the EU can work in any other EU country without restriction.

*LEGALLY YOU CAN ONLY DISMISS SOMEONE ON THE SPOT FOR GROSS MISCONDUCT*

**Firing** It's not easy to fire someone, even if they are bad at their job. Legally you can only dismiss someone on the spot for gross misconduct – stealing, fraud or the like. You can make someone redundant, but only if that is a genuine redundancy.

To fire someone, after the initial probationary period, without charges of unfair dismissal, you must give them the opportunity to improve, with appropriate training and support if needed. There might be some good reason for their apparent poor performance – medical or family problems, for example – and a sympathetic approach to resolution will win you not only the respect and loyalty of this employee, but have a beneficial effect on the rest of the workforce. It might be that the job you are asking them to perform is just not possible, or has in-built contradictions that need fixing.

All this needs dialogue, and that is what the law provides for. Typically you need to show that you have given the employee *two formal verbal warnings* and an additional *two formal written warnings*, each with discussion, for example with their line manager and someone from Human Resources. If action, such training, is suggested as a result of these discussions, you must be able to show that you took this action.

Only if there continues to be no improvement can you consider dismissal. Even then, you may be challenged to show to the employment tribunal that you have not unfairly dismissed the employee, so keeping accurate records is vital.

**Redundancy** A noted above, you can make someone redundant, but only if the job is actually ceasing. The employee is entitled to statutory redundancy payments, which typically in the UK are:

- half a week's pay for each year worked between 18th and 22nd birthday
- one week's pay for each year worked between 22nd and 41st birthday
- $1\frac{1}{2}$ week's pay for each year worked after 41st birthday.

These are subject to a maximum of 20 weeks' pay.

If the company is insolvent, there may be appropriate governmental guarantee schemes for these payments. They represent minimum amounts. The individual's contract or statement of work may contain additional benefits, as well as resolution of issues such as pension rights or stock option rights, etc.

‘IF THE COMPANY IS INSOLVENT, THERE MAY BE APPROPRIATE GOVERNMENTAL GUARANTEE SCHEMES FOR THESE PAYMENTS’

## Non-discrimination

It is always unacceptable, and in most jurisdictions it is illegal, to discriminate on the grounds of sex, race or disability (‘differently-abled’) where these would not prevent the job being done. However, there are unfortunately still some places in the world, and some traditional industries, where discrimination is still rife and in such cases, particular effort needs to be made to instill a non-discriminatory corporate culture and ethos.

The law also provides safeguards against sexual harassment in the workplace, which is taken particularly seriously in the US. Again, this is dependent on corporate culture and lead as much as anything. In any case it is a good idea to set up a friendly grievance procedure that allows sympathetic hearing and independent investigation of harassment complaints, before trouble occurs.

## Contracts

Almost every business transaction is governed by some form of contract, real or implied. Contracts are voluntarily making the agreements you want to make, and avoiding the commitments you don't want to accept.

For a contract to be valid there must be an offer and the acceptance of that offer. Offers (and their acceptance) need not be in writing, although there is an old saying that a verbal contract is only worth the paper it is written on, because of the difficulty of proof. A particular legal nicety is that you probably do not want to make an offer of a contract, for example on your website, as acceptance, and hence binding you to the terms, is then in control of the customer. You might want to refuse at that point, for example the goods might be out of stock., and, if a contract is formed, you cannot refuse. Rather, the terms and conditions should make it clear that what you offer is an offer to treat. The customer then accepts this, and it is then up to you to decide to conclude the transaction. Goods displayed in shop windows have been held to be offers to treat, in the same way.

‘OFFERS (AND THEIR ACCEPTANCE) NEED NOT BE IN WRITING’

Various terms are usually deemed to be incorporated into the contract, unless explicitly revoked, such as those defined by:

- Sale of Goods Act 1979
- Supply of Goods and Services Act 1982.

The law prescribes limits on contracts, for example in:

- Unfair Contract Terms Act 1977
- Unfair Terms in Consumer Contracts Regulations 1999 (**www.hmso.gov.uk/si/si1999/19992083.htm**).

Even if the terms are not written down, a contract is established if there is an actual or potential exchange of value, which includes promises. For example, in most business transactions there are presumptions of quality, safety, and so on. Some contracts have mandatory cooling-off periods, for example provided by the UK Consumer Credit Act 1974 and the EU Distance Selling Directive.

The law relating to contracts is complex, and if in any doubt get professional advice.

## Signatures

For a contract to be formed, there must be evidence that all parties agree. For contracts of significance, this is normally done with a signature, which is often witnessed by a second party. If evidence of the agreement is required, the witness to a signature may be called upon in court to testify that it was indeed the named person that signed the contract, and that the signee was not under duress or otherwise incapacitated.

A corporate contract is usually just signed by an authorized person. Physical corporate seals are no longer required in the UK, even for sealing. Where sealing is required, one or in some cases two authorized signatures will suffice. A seal may be needed for some foreign documents in overseas jurisdictions. The articles of association normally lay out the conditions under which the seal can be used, such as being used after a resolution of the Board approving the use of the seal. Of course, the individual signatory must be authorized to commit the company to whatever is agreed to, and part of the normal diligence in forming a contract would be to establish this fact.

## Internet issues

With the accelerating growth of e-commerce, the number of contracts formed over the internet is growing all the time. However, contracts and the signing of contracts conducted over the internet have particular difficulties, mostly connected with security, and these are proving a barrier to some users.

> ‘THE FIRST PROBLEM OF AN INTERNET CONTRACT IS HOW TO SHOW THAT A CONTRACT ACTUALLY HAS BEEN AGREED’

The first problem of an internet contract is how to show that a contract actually has been agreed and, if so, exactly what the terms of that actual or implied contract are. In order to show that the user has agreed to a contract, you need to demonstrate that the user has taken some action that consciously binds them to the terms and conditions of the contract.

For a paper contract, this is one of the functions of a signature, and witnesses to the signature can confirm that it was indeed the person stated that signed, and that they were not under duress and were of sound mind. However, what is written on the dotted line does not *have* to be a signature as such, and there is an acknowledged tradition of illiterate people 'making their mark', or the use of seals or 'chops' in Eastern traditions. At auctions, and in some trading pits, binding commitments are often made by hand signals or signs. Verbal contracts are by word of mouth. As long as the mark, chop or nod is acknowledged to signify assent, that is enough.

Electronically, it is rather harder to prove a party's agreement with a contract, since you need to demonstrate that the user has deliberately taken some action, such as pushing the 'accept' button or entering their password, and is fully aware of what they are agreeing to. You also need to show that it really is the user that is doing these things, and not some bug in the program, or a hacker. Thus, as a minimum you need:

- A way of displaying the terms and conditions to the user, together with confirmation of the transaction. You cannot force the user to read them all, but they must at least be aware of their presence, and have the opportunity to read them before agreeing to the transaction.
- A reliable authentication scheme to establish that it really is the user. There are various electronic schemes involving trusted third parties issuing electronic certificates, and more secure schemes that use some biometric measurement such as fingerprint or retina patterns used for closed systems. Schemes that involve physical tokens, such as smart cards, only establish the presence of the token, which might have been stolen or otherwise misused. Passwords are a common solution here, with the presumption that only the user (or their authorized representative) knows the password, although there must be some doubt whether the imposition of such a presumption on a consumer would survive the reasonableness tests under the Unfair Contracts Term Act and the more recent EU-based regulations. Passwords need to be of sufficient length to be difficult to guess, and they should also

be changed regularly. There is of course a compromise between degree of security and usability, and between risk and likely threat, which in turn is dependent on the value of the assets that are protected. The usual advice, for an everyday transaction of moderate value, is to use a password at least six characters in length, changed not less than monthly. You will need systems that can ensure that the password is transmitted in a secure fashion, such as using SSL encryption with a suitable key length, and that

— the password stays secret at the merchant's end,

— the password is not available, to techies or operators or help-desk staff scanning the logs or on some internal file,

— the software that handles it can be trusted not to disclose it.

- Secure audit logs that show that the binding action has occurred, showing for instance, that the password was correctly entered, in association with a particular transaction. These logs must be incapable of being tampered with, for example, by writing them to a write-once media such as a CD. They must then be kept securely. However, it's no good just writing the logs. There must also be a controlled way of reading and analyzing them. In the worst case scenario you may need to prove, in court, that a transaction occurred, and this could be in the face of hostile expert witnesses who will try and cast doubt on the integrity of you and your systems.

The next problem with an internet contract, is how to ascertain under which country's laws any disputes should be resolved, and subsequently in what courts they should be resolved. This problem arises from the global nature of the internet, which means that the buyer may be in one country, the seller in another, the ISP in a third (and routing the traffic through several others), with the banking and settlement system in yet other countries. Fortunately, this issue is one of the things that can be specified in the terms and conditions of the contract, using wording such as:

'This contract shall be governed by the laws of England, and any dispute shall be heard in the English courts to whose jurisdiction all parties hereby submit.'

Even then, collecting the judgment might be difficult against someone in another jurisdiction. Some countries (such as the EU) have agreements that recognize the judgments of another country, but it is by no means universal. Curiously, binding arbitration is much more widely recognized, and you might want to consider contract terms that resolve disputes by arbitration. Arbitration is also often a more cost-effective mechanism as well.

In practice, since credit cards (or a banking system) are used for the majority of payments of internet transactions, it is the rules of the banks and the credit card company, rather than that of the courts, that prevail and set the ground rules. Such companies act as a universal currency and decide who gets paid and who does not. Credit cards are the engine that enables online trading to work: they act as the trusted third party, and provide the guarantees of payment and of delivery. However, credit card payments can be cancelled unless there is proof that the customer really did order the goods, and they may not guarantee payment where the customer is not present and the merchant cannot produce a voucher with the customer's signature. Read the fine print on your credit card merchant agreement.

“YOU MIGHT WANT TO CONSIDER CONTRACT TERMS THAT RESOLVE DISPUTES BY ARBITRATION”

Of course, the transaction still has to be legal, and this can also be a problem for internet transactions, as standards of legality vary from country to country. In particular, the rules and regulatory agency for financial instruments such as stocks and shares, bonds and life insurance can vary widely from country to country, and something that may be perfectly legal in one country may not be so in another. For example, shares in a UK company, regulated and traded in the UK on a UK exchange, may legally be owned and traded by a US citizen. However the same shares may not be advertised for sale in the US, including on a website, without the authority of the US Securities and Exchange Commission.

Gambling, guns, fireworks, drugs including medical prescription drugs, pornography, wine and spirits, Cuban cigars and other, mostly pleasurable things, also have different regulatory regimes in different countries, which may affect delivery and fulfilment of the contract. Some contracts, like credit agreements and property transactions have mandatory cooling-off periods in some jurisdictions.

Because of the global reach of an advertisement on a web page, it may be wise to include words to the effect of 'offer not valid where prohibited' in the disclaimers on the site, and to positively screen out access from, or shipments to, addresses in countries which are known to have laws which could be hostile to your product or service. Except for extreme cases of gross fraud or child pornography, prosecutions for infringement of a country's laws by e-commerce have so far been rare, although it is better to be safe than sorry!

**Legal changes** Many jurisdiction's laws are slowly changing to accommodate internet trading and activity. For instance, specific forms of electronic signature are now legal in the UK and US even for documents where a manual signature was previously mandatory.

However, in some countries, the encryption that underlies privacy agreements and signatures is illegal, or is counted as a controlled armament. In most countries, including the US and UK, encryption is permitted, but the government (and employers) retain the right to eavesdrop, and to impose censorship. Clearly, this situation creates a strong potential for our civil liberties to be severely compromised, and it is incumbent on all of us to fight for our rights. Indifference to these fundamental rights to privacy will allow governments to take them away.

## Exercises

**one** Go and visit a law court. Resolve not to be there as a litigant.

**two** Write briefing documents to enable your (future) legal adviser to draw up:

- changes to the company's memorandum and articles of association
- shareholder's agreement

— director's agreement
— initial client contracts.

Focus on what might be different and the special things that you might need, rather than the standard terms and conditions.

## Further reading

Keenan, D. and Riches, S. (1998) *Business Law*. FT Prentice Hall

Renton, T. and Watkinson, J. (2001) *Company Director's Guide*. Kogan Page in association with Institute of Directors

**www.parliament.the-stationery-office.co.uk/**
(Legislation.)

**www.venables.co.uk** Delia Venables
(About legal sources and law firms on the web.)

**www.legalpulse.com**
(A free service for start-ups.)

**www.emplaw.co.uk/free/** Disc Law Publishing Ltd
(Publishing company specializing in legal texts.)

### *Legal practices*

**www.collyer-bristow.co.uk/news-publications/summary-of-directors-duties1.htm** Collyer-Bristow, London

**www.launchpadonline.co.uk** Mills & Reeve, Cambridge

### *Example documents*

**www.digitalpeople.org/resources.html** Digital People

# 8 Setting up and recruitment

‘One’s life is frittered away by detail.’

(Henry Thoreau, 1817–62)

THIS CHAPTER IS MOSTLY ABOUT SETTING UP A company, and finding the staff to work in it. Setting up requires lots of detail things to be done, before you can get on with the business of whatever it was you set the company up to do.

## Setting up

The easiest way to form a company is to purchase a ready-made one either via your lawyer or directly from a company formation agent such as Jordans, at a cost of a few hundred pounds. For this you get a complete package of company documents, a company seal and everything else needed. For instant companies, they have a whole series of off-the-shelf ones with silly names. Incidentally, that is how one of my companies ended up with the name Topexpress Ltd, as we never got round to changing it.

Instructions are included to enable you to change the directors and the company secretary to the candidates of your choice. A private company must have at least one director and a company secretary. A public company needs at least two directors and a company secretary. You can also change the company name if need be, and the memorandum and articles of association, to reflect your company's needs. You are required to file the appropriate forms with the relevant authorities to register your company (Company's House in the UK).

Next, you need to hold the initial meeting of directors, pass the initial resolutions and appoint auditors and bankers. The kit includes a minute book in which you must enter minutes of the formal resolutions of the initial meeting. There is a special form for this. You will also get a share register, and the company secretary must issue share certificates and record them in the register. Now you have to get down to business, or at least planning the business. There are masses of details to attend to.

## Premises

One of the first things to consider are your premises; you can't work out of your bedroom forever. If you are a new company and only require small amounts of space, serviced offices may be a good choice. Serviced offices provide a complete package – premises, furniture, heat and light, telephones – and a receptionist to answer them! Many also come complete with photocopy and fax facilities, as well as things like secretarial services and meeting rooms if required. Serviced offices are often available at short notice, with flexible leases. Some are designed explicitly as incubators for high-tech companies, and can include advisory services. Others, typically associated with universities or other institutions, have facilities designed to cater for emerging technologies, such as laboratories or multimedia studios. In such environments, you will be next door to other bright new companies, and the value of the social interaction, contacts and possibly competition that this can provide should not be underestimated.

However, although serviced offices are a quick and flexible way to launch your company, they are also numbingly expensive, and so are not a good long term solution. In the long term you will probably want to lease your own premises, which is a much better use of your precious capital than purchasing. However, once you have left serviced accommodation, you will have to provide for yourself all the services that were previously provided for you. By the time you get your own premises however, you will hopefully be big enough to employ someone to sort this out for you.

Beware of fancy office outfitters. Contrary to popular belief, there is no need to get fancy furniture, or to spend lots of money on the reception area. Far from giving a good impression, it will give the impression that you are spendthrifts, and not to be trusted. Second-hand furniture is often good

## UK company types

These definitions are based on UK law, but most jurisdictions have equivalents.

The main entity types are:

- *Sole trader*. A one-person band. A sole trader is solely responsible for all their actions, and all their debts.
- *Partnership*. People who have agreed to work together, governed by the terms of their partnership agreement. Partners have joint and individual responsibility for their actions and debts, although recent changes to the law allow partnerships to limit their liability. Professional practices, such as accountant and law practices, are often partnerships. If you are contemplating a partnership make sure you get the terms of the partnership agreed while you are all still friends.
- *Limited private company (Ltd)*. A private company, but one with a limited liability to pay its debts. If it all goes wrong the directors are only liable for the amount of their original investment. However, most banks and larger trade creditors will want additional and personal guarantees before doing business. Governed by their memorandum and Articles of Association, and Table A of the Companies Act 1985.
- *Public limited company (plc)*. A limited private company, but one in which shares in the company may be sold and traded by the general public, via recognized financial intermediaries. There are further additional and onerous reporting requirements.
- *Listed company*. A public limited company whose shares have been listed on a recognized public stock exchange. The exchange will impose additional reporting and trading restrictions.
- *Special cases (e.g. trusts, societies)*. There are various special cases, such as charitable trusts, building and other societies, banks that have additonal requirements, and special corporate entities.

Most start-ups will be a limited private company.

value, and there are many companies that specialize in selling second-hand furniture to new companies, and then buying it back from them as they move upmarket.

## Telecommunications

You need to sort out a phone number and internet access as soon as possible, but these services will ideally be part of a more comprehensive communications infrastructure plan. At the very least, you should add some some sort of answering service to this list, to make sure that calls are not missed if you are not there, or even better, a redirection service, to your cell-phone, for example.

As you grow, your communications system, and the facilities that it can provide should grow with you. A virtual private exchange such as the Centrex system may be suitable, and will not eat all of your capital. Your telecom provider will be able to advise you on such matters.

As soon as you have set up internet access, you should also get a domain name, and start your own website. Following this step is the requirement for your own intranet, firewall, web and mail servers. Shared electronic diary systems, such as Microsoft Outlook, can also be of immense benefit to your communications infrastructure, as they allow other people to check your diary to find out where you are, and you can also arrange meetings easily. Of course, if you start to spend lots of time out of the office, you will also require secure external access to your internal communications systems, enabling people to work from home or on the move.

## Stationery and corporate image

It is important that you always present your company well, and consistently, and a strong corporate image can help you to achieve this. In some companies, corporate image can be all-pervading, with every last detail reflecting it. Unless you are a very image-conscious company, you will initially only need to consider the company name and logo, and general style of these. It's probably worthwhile getting a company logo professionally designed, as you will have to live with it for some time. However, straightforward typography can be stylish, classic, and give a reassuring image. To paraphrase a quote from a fictional TV producer, in Anthony Jay's *Yes Minister*, 'A jazzy modern image means that you have nothing much new to offer; classic traditional styling conceals a radical message.'

Once you have sorted out your logo, you should get some stationery printed, although you can manage for a while with laser printed notepaper. You will also want to get compliment slips and visiting cards for everybody.

The letterhead must show the registered office address, the country of registration and the company's registered number, such as 'Registered in England No. 1233434'. It is no longer necessary to show the names of the company's directors, and inadvisable, as they may change.

## Financial control systems

You need to get some sort of accounting system set up straight away, as soon as you have a bank account. It is a statutory duty to keep proper books of accounts, and chucking your receipts in an old shoebox won't do for long. You need to get organized before it gets on top of you. You might consider employing a part-time bookkeeper.

Make sure that you get control of the chequebook, and set up a proper purchasing system. This should have limits on what anyone, including you, can spend without agreement. How much you authorize on a single signature depends on the size and spending patterns of your company. Too much is a temptation, if not for fraud then for unconsidered purchases. Too little and it becomes onerous. An example would be a spending limit of £500 on a single signature, and £5,000 on two, with anything greater needing a full Board resolution. Others require two signatures for anything up to £1,000, and Board authorization above that. Although you might be able to trust your partner, you will be employing people who may not have quite the same ideas. It is much better to establish some control at the beginning, before the money starts to walk with people buying themselves things such as new notebook computers on a whim. Similarly you need to get an expenses system in place.

To go with your order entry and purchase control system, you need an asset and inventory system, so you know what you own and what you don't. You should mark major items, like computers, so that you can identify them, and your auditors can count them. No doubt the police and your insurance company will appreciate anti-theft invisible UV markings as well.

So now it's time to blow the dust off the budget and try and make it work. Are the assumptions still true? Have you already spent more than you planned? If so, how does that affect the cash flow? Fix dates for budget review meetings in the new diary system, and make sure that the accounting system generates the information you need to be able to compare your actual performance with the budget.

It is a statutory duty to have employee liability insurance, and to display the certificate.

You may need buildings insurance for the premises, and insurance for the contents. Your investors may want you to take out key man insurance – a sort of life insurance in favour of the company. Pensions and health insurance might form part of the remuneration package and are discussed in the chapter on people. Beware of heavy administration charges!

## Staff and recruitment

You need to plan any recruitment. What you do at the start sets the culture for the company. Now is the time to get into good habits. Weekly team meetings ('Monday prayers'), regular progress and budget reviews, monthly Board meetings, and regular beer or curry nights, or whatever you do to get the informal communications flowing, should all go into your newly set-up electronic diaries.

## Start work

You need to set up relationships and accounts with key suppliers, and purchase initial kit, and computers.

You have to build whatever you said you were going to. We discussed project plans in Chapter 10, but now is a good time to review and to add the next level of detail.

We discuss quality plans in Chapter 12, but you need to get started now. Quality is not a bolt-on option, but needs to be built in from the beginning. How else will you tell if what you have built works?

## Start marketing

We discussed marketing in Chapter 2, but it is another thing that needs to get started right away. You need to make those plans, and start making those phone calls and building contacts.

## Review

Now you are up and running. How does it feel? How does it compare to your plan, and what does that say for the future?

# Recruitment

This section covers recruitment, remuneration and interviews. Legal aspects, such as equal opportunities and employment law, are covered in Chapter 7.

'I NOT ONLY USE ALL THE BRAINS I HAVE, BUT ALL I CAN BORROW'

(WOODROW WILSON, 1856–1924)

Recruitment is, as a rule, expensive, difficult and time-consuming. However, if you are able to find the right people, it is well worth it. Before you begin, spend some time figuring out the kind of person you want, not just at the level of 'good verbal and written communications', but more who would be the ideal person for this job. Write a half-page pen-portrait. What background would they come from? What experience would they have? Where would they expect their career to go? Why would they want to join your team? What would they be earning at present, and what would you have to do to attract them? It might not be just salary, but better prospects, or a good options package might tempt the ideal candidate.

By far the best way to find new people is through personal recommendation, which is quick, cheap, and low risk – at least you should know what you are getting! Some companies value this method so highly that they run bounty schemes to reward staff who introduce new employees. In such cases, the person who made the introduction is typically paid some months after the engagement of the new employee, such as at the end of their probation period.

If your own staff cannot recommend or find new recruits, you can employ an agency to do this for you. However, agencies are typically quite expensive, charging 15–20 per cent of first year salary. Their effectiveness can vary greatly, although they are usually most effective for recruiting contract and temporary staff.

For more senior posts, headhunters are effective, and will minimize publicity, if this is an issue. However, they are very expensive, and again their effectiveness varies, as does the breadth of their address books. Pick your headhunter as carefully as you would the candidate you are hoping to recruit.

Advertising can be a good alternative, or addition, to the use of agencies or headhunters, and need not cost you much. Indeed, advertisements placed on internet sites, or in the relevant technical or local newsgroup, can often be low cost and effective. Advertising in the local or national press is more expensive, and will require careful wording for maximum effectiveness. In these days of skill shortages, advertising is more than just informing people that you are recruiting; part of the purpose of advertising is to try to sell the company to potential employees. Like any advertising, it is useless unless you have a means of collecting, filtering, grading and replying to the responses.

# Interviews

The three main reasons for interviews are:

- recruitment
- appraisal
- discipline.

We examine the recruitment interview first, and then look at the others.

## Why

Why have an interview at all? After all, 15 minutes of contact cannot add that much to information obtained from the candidate's CV and references. None the less, the interview is an important part of the selection process and can assist in:

- Learning more about the candidate. You can verify information, and fill in any gaps.
- Comparing the candidate with the job requirements. You will be able to make a better assessment of how the candidate might fit in with the other members of the team.
- Selling the organization and the job to the candidate. Do not presume that if someone turns up for an interview they definitely want the job!

## Shortlisting

If you receive more applications for a particular post than you want or have time to interview, you will need to shortlist. Shortlisting, and the administration involved with arranging interviews, can sometimes take up much more time than the actual interviews themselves, and it is important that you have identified how you will deal with the process before you advertise. Try to make sure that you respond to both the successful and unsuccessful candidates, and that you get back to them as soon as possible. If there is too much time left between the application deadline and interview dates, candidates may lose interest or find themselves another job.

As with the rest of the recruitment process, it is very important that you can account for your decisions when shortlisting, as if not, you could fall foul of equal opportunities laws. One way in which to choose candidates fairly is to draft a list of the main criteria for the job, and tick off which candidates match most criteria. These candidates will obviously be those that are invited to attend an interview.

## Preparation

So, after much time and effort, you have generated a list of people who are interested in working for you, and have whittled them down to the few candidates who might be suitable. Think of them as sales leads, and set up the meeting accordingly. Preparation is essential to make the most of an interview, and should mean that you ask all of the questions that you intend to ask, and extract all of the information that you need.

There are some important points to consider when preparing for an interview:

**Who should be present, and who needs to know about the interview?** It is good to have more than one person interviewing as it can be helpful to compare your thoughts with colleagues. It may be useful to involve someone from Human Resources, although much more important is someone from the team that the candidate could be working in. You should also inform and possibly involve the internal job sponsor, that is, the person to whom the candidate will report. Make sure that all of the relevant papers, including CV, job description, arrangements and report form (see later), are circulated to everyone involved, well in advance of the interview. Also, ensure that all involved are clear on the format and structure of the interview.

**What format should the interview take?** Think about how the candidate will be greeted on arrival, and who should do this. Where will the interview be held? Ideally, you will require a quiet space where you will not be disturbed. Will you provide refreshments? Will you take the candidate on a tour of the office or plant, and should the candidate meet existing staff and, if so, whom?

**What do you want to get out of the interview; what questions do you need to ask?** Decide and note down what questions you want to ask. Think carefully about who should ask what questions, and how many questions you will need.

**What do you need to tell the candidate?** Prepare an outline of the company and the job – this will may be something that is sent to the candidate prior to interview. Think about what the candidate may need to know, about the job, the company, the location, and the future, and make sure that you can answer such questions.

**How should you assess the interview?** Generate a report form so that everyone interviewing the candidate can compare notes in a common format. As with the shortlisting process, you must be able to fairly account for your recruitment choices, and report forms can be a useful form of documentation if you are required to prove the equitability of your decision.

## Conducting the interview

Interviewing is a skill and an art. The best interviewers are good listeners, and are able to put the candidate at their ease, and not impose too much of themselves or their own biases into the conversation. Examples of how not to interview abound, such as the one where the interviewers spent the whole time arguing among themselves, with the candidate scarcely uttering a word. Incidentally, he got the job . . .

Some common examples of bad practice include:

- *Pre-conceived ideas*. People inevitably bring some bias to an interview. However, it is very important that candidates who could be ideal are not overlooked, simply because they do not fit with the preconceived idea. Conversely, do not fall into the trap of employing a weak candidate just because they appear on the surface to be the right type.
- *Only remembering the last candidate*. If you are interviewing several people, it is hard to remember them all, and the details can get

confused. A report form filled in after each candidate helps. Some companies even take a quick snapshot of candidates, with a digital camera.

- *Not letting the candidate get a word in edgeways*. It is the candidate who is being interviewed, not you. Not listening to the candidate is the biggest single fault in interviews. Also make sure that you pay attention to the candidate's tone and body language – do they portray a different picture from what they say? Are they unusually nervous or over-confident?
- *Asking closed questions*. Do not ask the candidate questions which have only yes or no answers. Instead, try to ask questions that will encourage discussion. An interview is not an adversarial process. You should attempt to create rapport with the candidate – smile, make eye contact, and don't just mumble to the table. Overcome awkward pauses by summarizing the points so far, or reflecting on the last answer. Long silences are stressful but can be used deliberately, although being interviewed is in itself stressful.
- *Asking personal or politically-incorrect questions.* Some questions, if they do not directly relate to the performance of the job, should not be asked at all, for example 'Are you married?', or worse 'Are you pregnant?'

## The interview itself

Here are some pointers for the possible format of an interview.

**Open** Start by welcoming the candidate, and introducing the people present. Check that the candidate is who you think they are (confusion can easily arise) and summarize the job and the recruitment process. Try and put the candidate at ease, for example by making small talk about their journey. Then talk about the job and the company.

**Discussion** Maintain a moderate pace. Allow the candidate to answer fully, but keep the momentum going. Ask open, not closed questions, that is questions that allow discussion, not a simple yes or no. For example, instead of asking 'How long were you at Bloggs Ltd?', ask 'Why did you leave Bloggs Ltd last year?'

Ask situational questions, such as how would the candidate deal with some work-related, or hypothetical situation? Probe past experience, and check anything fuzzy. Check any gaps in the CV – maybe they were in prison or have something they want to hide – and check any unusually short job durations. If the candidate seems unusually stressed about some point, try and find out why.

**Close** Check your plan – have you covered everything? Ask if the candidate has any questions, or needs to cover anything else. Explain the next stage, and make a commitment as to timescale ('We will write to you next week').

Check if the candidate is still interested – you may have put them off, or they may have other job offers. Remember, in a culture of skill shortages, the candidate will be in a very strong position, and may be interviewing you, as much as you interviewing them.

Complete any housekeeping, such as expense forms for travel re-imbursement, and then complete the process by, for instance, passing them on to the next person, taking them on a tour of the plant or seeing them back to reception.

## After the interview

**Make the decision** Does the candidate have the right skills? Do they have personal qualities appropriate to the post? Will they fit into the culture and the rest of the team? Is this candidate the best compared to the rest?

Once the decision is made, check the CV again. There is unfortunately a small proportion of people who are fraudsters and make up their qualifications and references. Always check the references, and follow up by telephoning any ambiguous replies ('You will be lucky if you can get them to work for you' or 'My colleagues and I hope he finds a new job as soon as possible'). Get independent confirmation if in any doubt.

## From the candidate's perspective

Interviews are crazy. You've got maybe 15 minutes to sell yourself, and check out the organization that you are potentially going to devote your energies to for a good proportion of your working life. Do your homework. Find out as much as you can about the company or organization beforehand. If you can't be bothered, why should they bother with you? Check their website. What are their main achievements, and their main competitors? Talk to people in the industry. Where is the company going? What do you see as your role, now and in five years' time?

At the interview, above all, be youself. It's hard to maintain a front, and your interviewers will probably see through it. Try not to be nervous – they are ordinary people, and have been through such processes themselves. As the old salesman said, it doesn't matter who they are, they still have seven pounds of hot manure inside them, except that he used a shorter word than manure.

One thing that can be very persuasive in an interview is a high level of energy. Be enthusiastic, and talk about your achievements, whether they are directly related or not. Remember the STAR technique detailed below, which can be used as a guide for providing strong answers. The technique can be utilized to discuss effectively a wide range of experiences, from classroom projects to work situations and extracurricular leadership activities.

**S**ituation – discuss a situation or problem which you have encountered.

**Follow-up** Once the decision has been made, get the job offer out as soon as possible. A formal contract or statement of work needs to be issued within two months of starting, or within one month for overseas employment.

Set up whatever induction process you need. What happens when they turn up for work? At the very least, someone should act as a guide and mentor for the new starter, although many companies have formal induction schemes.

Task – discuss the task that the situation required, or your ideas for resolving the problem.

Action – explain the actions which you took, and obstacles that you had to overcome.

Results – highlight the outcomes, and what goals you achieved.

Think of questions you might like to ask the company, if nothing else to show interest and commitment.

Examples might be:

- What are the best and worst aspects of the company?
- If I join the company, what are my prospects – where will I be in three years' time?

Other examples of questions are given on the website and in the Further Reading section below.

Finally, think of reasons why you want this job and why you want to work for this company, other than the money. If you can answer these convincingly, you are halfway there.

If, at the end of it all you don't get the job, don't be too despondent. There may be many reasons why you weren't successful, not least internal company politics. Don't worry – another opportunity will be along soon, and at least you will have had some practice. It can sometimes be helpful to ask for feedback on an interview – sometimes people are happy to discuss this, or they may prefer to write to you.

## Appraisal interviews

It is considered good practice for every employee to meet with their line manager on a one-to-one basis (or, in the case of the CEO, to meet with the Board) for an appraisal at regular intervals, typically quarterly or every six months. Although formal, these are not adversarial meetings, but help gain common ground on reviewing the previous period and planning for the next one. Appraisal meetings enable team members to get a clear idea of how they are doing, and can identify where they might need further support or training. They are also an opportunity to set objectives for the next period, and to plan the team member's personal career and growth.

The interview is typically centred on the appraisal form, which is kept with the employee's personal records. A typical form would include:

- date, name, job title, assessor
- self-assessment – completed by the candidate
- assessor or line manager assessment
- key objectives – jointly agreed
- development plan – jointly agreed
- actions – jointly agreed
- follow-up dates, meetings.

## Discipline interviews

In some respects, discipline interviews are similar to appraisals, as they are a way of seeking solutions to problems. Discipline interviews form part of a disciplinary investigation, but it should be stressed that they are not a kangaroo court, and should be focused on overcoming any problems. Refer to the section on employment law in the last chapter. Such meetings should not be conducted by a line manager on their own, and should involve, for instance, a representative of the company's Human Resource section who can act as an impartial witness in case of subsequent legal action. The candidate may also wish to have a work colleague, or a friend, or a member of their union present in support.

The procedure of a discipline interview would typically follow this format:

- State the accusation and the result of any investigation.
- Ask the employee to state their case, and explain any mitigating circumstances.
- Announce formally the findings, warnings, etc.
- Work on a resolution, and support needed, etc.
- State follow-up actions and dates.

The results need to be formally recorded. If termination of employment results, the normal HR processes need to swing into action, assisting with relocation, recovering any company property, transferring pension rights, etc.

## Remuneration packages

Remuneration is more than just salary. It can involve stock options, cars, profit-share schemes, bonus schemes, pensions, travel and other perks. For some jobs, commission payments are important, or there may be particular tax-planning issues.

Salaries are, however, a principle component of remuneration packages. Don't stint – as the saying goes, 'Pay peanuts and get monkeys'.

Although as a start-up you will tend to pay whatever you have to for the right people, even with a small number of people this can lead to anomalies between the salaries of people doing comparatively similar jobs, and this in turn leads to discontent. It might be worthwhile to give some thought to getting a rough salary structure in place from the outset, and using it to guide new salaries and salary reviews.

### Stock option schemes

Stock options allow employees to participate in the capital growth of the company. Essentially, they provide the employee (the option holder) with the option to buy shares in the company at a fixed price (the strike price) at a future date (sometimes after a certain date). For a high-growth company, hopefully the value of the shares in the future will be much more than the strike price.

However, stock options are high risk – they can be much more valuable than the total salary, or work out to be absolutely worthless. Indeed, they have no value at all until there is a market for the shares, for example following either the company going public or being acquired. The value depends not only on the performance of the company, but also on the market. Stock options are most suitable for key staff in high-growth companies, as they encourage staff to optimize the company's share price, and align their interests with those of the investors. They can also be used as a means to retain staff, as well as an aid to staff recruitment and motivation.

For young companies, they are paper remunerations, not using the company's precious cash, and can be tax efficient.

The design of stock option schemes is complex, and needs to balance the advantages to the company and to the employee. An example scheme would involve a four-year accrual with a one-year cliff. The one-year cliff means that the employee gets nothing for the first year, but if they are still employed at the end of that year they will get the first year's options. Thus, assuming continuous employment:

- During the first year no options can be redeemed.
- After one year 25 per cent of the options granted can be redeemed.
- After two years 50 per cent can be redeemed.
- After three years 75 per cent can be redeemed.
- After four years all can be redeemed.

For non-quoted companies, that is companies whose shares are not traded on a public exchange, the strike price would typically be either par, that is the nominal face value of the shares, or the effective value of shares at the last investment round. For quoted companies whose shares are publicly traded, the strike price is typically the average market value of the shares over the previous 30 days prior to the grant of the option. If the grant price is very different from the market price at the time of grant it may attract tax on the difference.

In most jurisdictions stock options only become liable for tax when they are exercised. Any scheme should be structured to take advantage of the tax benefits available, which can be substantial.

## Profit share and other bonus schemes

Bonus schemes are only useful motivators if the person's performance can affect the outcome. While stock options and the like work well for key senior employees, cash-based schemes are more relevant for others. For younger staff, cash in hand is more useful than a promise of jam tomorrow, especially if tomorrow never comes. Many bonus schemes are based on 'profit sharing', which typically involves distributing 15 per cent of pre-tax profits equally between every employee, on a pro-rata basis for the period. Some companies operate bonus schemes based on individual appraisals and performance of the division or project team, where such indicators can be meaningfully measured.

## Other remunerations

**Pension schemes** Pensions are complex and technical, and independent specialist advice should be sought to maximize the tax benefits. With increasing staff mobility, contributions are more normally paid into the individual's personal pension fund, rather than consolidated into a company pension fund.

### Commissions

For sales-related jobs, commissions form an important part of the remuneration package. A typical commission might be as much as ten per cent of the sale price, paid on contract signature. For large contracts this can be a significant amount.

### Cars

A company car used to be a tax-efficient form of remuneration, but this is no longer the case in most countries. Also, because they tend to be used as status markers, company cars (and similarly reserved parking places) often generate more aggravation, and demand more management time, than is reasonable. It is better to have a generous mileage allowance scheme, encouraging people to use their own cars, and operate parking on a first-come first-served basis.

## Exercises

**one** Survey local premises in your area. What are the local planning laws about setting up at home?

**two** Draw up an action plan for what needs to be done to set up. What are the highest priorities, and what can be left until later?

**three** Practice recruitment and appraisal interviews on each other. Which of the Belbin worktypes (www.belbin.com ) would you assess the candidate as, and why?

**four** Draw up job descriptions for the first five employees, and what their responsibilities will be. Write a brief pen-portrait of the ideal person for each role.

## Further reading

### *Setting up*

Jay, R. (2000) *Fast Thinking Selection Interviews*. FT Prentice Hall

Jay, R. (2001) *Fast Thinking Manager's Manual*. FT Prentice Hall (Covers interviewing, appraising, dealing with difficult people, and so on.)

**www.jordans.co.uk/**

(Jordans is the UK's leading provider of corporate professional services. Jordans' services include company formations and company secretarial services, software solutions, trademark and domain name registration, company searches and conveyancing services.)

**www.companies-house.gov.uk**
(The official organization for the statutory registration and provision of company information.)

**www.hmso.gov.uk** Her Majesty's Stationery Office
(The source for all government publications.)

## *Recruitment*

Smalley, L.R. (1998) *Interviewing and Selecting High Performers*. Kogan Page

**www.virginia.edu/~career/handouts/interview.html**
(Useful guidelines on being interviewed.)

**www.businessbureau-uk.co.uk/**
(Information resource for small businesses.)

# People, projects and products

# 9 Managing people

‘There go my people. I must follow them, for I am their leader.’

(M. Gandhi, quoting Alexandre Ledru-Rollin: ‘Eh! Je suis leur chef, il fallait bien les suivre’)

THIS CHAPTER IS ABOUT PEOPLE, OR 'HUMAN RESOURCES' as they are now called. Recruiting and managing people is a huge task and one that could fill a whole book in itself – in fact it has filled hundreds of them, which is why I'm not going into detail here – but it is crucially important to manage your staff well to get the best from them. So, if you suspect you might not be the most naturally gifted people person you should either find somebody who is to work with – or get a good book and learn fast.

The main areas that it covers are:

- management theories and motivating factors
- groups and teams.

People are your greatest asset. This chapter is about persuading them to help you achieve your goals. The main theme of this chapter is that people are the essence of your company. Modern management concentrates on the people and their relationships, not the tasks they do.

## Management theories

In 1960 Douglas McGregor published *The Human Side of Enterprise* (McGraw-Hill, 1960), a seminal text on the theory of management, in which the old and the new types of management are compared.

McGregor developed two company models to represent the old and the new management structures, which he called Theory X and Theory Y:

- Theory X describes a company with a traditional hierarchical structure.
- Theory Y describes a company with a people-oriented structure.

We examine each in more detail.

## Classical management theories

In order to understand modern management, it is helpful to look first at the classical theories.

Traditional hierarchical management theories were evolved in the era of the smoke stack, at the end of the 19th century. They focus on the structure of the organization, and the reduction of tasks into simple elements, leading to work that is often boring and repetitive. This structure was devised for a mostly unskilled labour force, and assumed that individuals were primarily motivated by pay and by the stick. As a result, punishments were often threatened if orders were not obeyed, and the whole culture was autocratic. One exponent of this theory at the end of the 19th century, was F.W. Taylor.

In Taylor's day, a manager would have had very little contact with the day-to-day work of the factory. The work would have been divided among small, semi-independent gangs, with a foreman in charge. Each gang used what tools they had or could get, and developed their own style of labour. As a result, workmanship was often shoddy, inefficient and poorly organized. One of Taylor's goals was to increase efficiency, and hence both the pay of the workers and the profit of the enterprise, by finding the best practice and then teaching this to the other teams.

In 1911, Taylor wrote *Scientific Management* in which he coined the term 'time study'. Taylor argued that all knowledge should be systematized, work standardized, with every task broken down into its smallest components, and each component optimized. By these methods, he reduced the number of labourers at the Bethlehem Steel works from 500 to 140, by methods such as introducing different size shovels for each job. In one particular example, he taught a (famously named) worker called Schmidt to load pig-iron all day by taking regular rest periods. At the time, Taylor was considered humanitarian for improving work conditions, and because in this instance he actually recorded the man's name.

The main assumption was that people don't want to work, and therefore they have to be made to do so, which can be achieved through coercion, control and threats.

F.W. Taylor's dictums were:

- Record all knowledge.
- Scientific selection and progressive development of workforce.
- Work for maximum output.

Applying his principles to the organization, his framework was:

- clear delineation of authority and responsibility
- separation of planning from operations
- incentive schemes for workers
- management by exception
- task specialization.

Frank and Lilian Gilbreth worked with F.W. Taylor, and took his principles further. They introduced strict 'motion study' and standardization of work elements. They advocated piece rates. As was the spirit of the times, they treated workers almost as automatons, and their innovations had the effect of further de-humanizing the workplace.

Henri Gantt was another employee of Taylor's and gave his name to the Gantt chart, which we shall discuss in Chapter 10. Gantt replaced piece rate with a day rate plus a bonus and, in contrast to the Gilbreths, was known for his humanizing influence on management theories, emphasizing the need for good conditions for the worker.

## Theory X

Theory X is based on the assumption that people don't want to work, and that they have to be made to do so, through coercion, control and threats. The primary motivation of employees in Theory X companies is pay, and the stick.

This inevitably leads to an organization with a hierarchical structure, with status demarcations between management and workers, creating a 'them and us' culture. Everyone has a clearly defined, task-oriented role, with little job flexibility. Jobs tend to be repetitive, and most management effort is spent on fixing things when they go wrong ('management by exception').

Companies with Theory X structures are typified by poor internal communications, and are slow to change or adapt to changing circumstances or markets.

## Theory Y

Theory Y is based on the assumption that people naturally want to work, but are prevented from doing so by the organization. The primary motivation of people in Theory Y companies is the carrot or reward.

‘PEOPLE NATURALLY WANT TO WORK, BUT ARE PREVENTED FROM DOING SO BY THE ORGANIZATION’

Given a supportive environment, people will seek and accept responsibility. They will also control their own work when they understand and are committed to common objectives. Theory Y companies put people at the centre. They consider individuals as individuals, with social and motivational needs, not just units of work. They typically have a flat management structure which is participative and inclusive (such as the matrix structure discussed below), with flexible work teams, and good internal communications. They have little, if any, demarcation between workers and management. Status symbols, where they exist, are likely to be based on merit and performance, rather than role. Such an organization adapts well to rapid change, and virtually every modern company aspires to follow this model. Needless to say, nearly all successful knowledge-based companies follow Theory Y.

## Maslow's hierarchy of needs

McGregor's ideas were influenced by a theory developed by Abraham Maslow in the 1950s. This theory stated that a hierarchy of human needs generates a person's primary motivation. Maslow arranged needs in levels, and argued that as soon as the lower level is satisfied, the next level emerges and demands satisfaction. Only unsatisfied needs motivate behaviour. As a crude example, if you are drowning, then the fact that you are also hungry is, for the time being, ignored. However, as soon as you are rescued, food becomes a priority.

Maslow characterized the levels as:

- Physiological
  - air to breathe
  - drink
  - food
  - warmth.
- Safety and security
  - protection from danger
  - health
  - economic security.
- Social and affection
  - give and receive friendship
  - belong, be accepted, part of a team.

- Esteem and status
  - recognized status
  - peer esteem
  - external recognition.
- 'Self-Actualization'
  - realization of goals
  - self-awareness
  - growth.

This hierarchy is, following Maslow, often drawn inside a triangle.

Following this rule, a worker's disaffection with work is due to poor job design, managerial behaviour and too few opportunities for job satisfaction, rather than an intrinsic fault in the worker. It is interesting to note that pay is not a direct issue in Maslow's hierarchy, but relates indirectly to economic security and to status.

‘INDEED, IN DEVELOPED SOCIETIES, WHERE PHYSIOLOGICAL NEEDS AND ECONOMIC SECURITY ARE NOT AN ISSUE, IT CAN BE ARGUED THAT KUDOS IS NOW THE PRIME MOTIVATOR FOR WORK’

Indeed, in developed societies, where physiological needs and economic security are not an issue, it can be argued that kudos is now the prime motivator for work. An example of where this theory holds true is in the Linux development community where people contribute, not for financial reward, but for kudos and status. Curiously, in these communities, people are persuaded to do the boring grunt work, such as maintenance tasks, by making these high-status jobs.

It is therefore essential that a job can satisfy an employee's needs and aspirations, if you want them to stay. However, implementation in practice may be hard, and much depends on the company culture. If a person's social and status needs are not met from within the company, they will try to satisfy those needs elsewhere, in, for example, some absorbing hobby or external pursuit which may well detract from the effort that they put into their job.

Most high-tech firms actively manage the physical working environment, aiming to satisfy 'physiological' as well as 'safety and security' needs, providing for example, a continuous supply of coffee, ergonomic seating and adequate and well-lit work areas, all conforming to the health and safety regulations. Some companies go even further, such as Microsoft, which provides a separate office for each employee, rather than just a pod in an open plan office. Kitchens on each floor provide a wide range of juices and snacks, and 'play areas' encourage employees to take work breaks, without straying too far from their desks. In some companies, fresh doughnuts are provided to encourage early starters, and freezers packed with ready meals, accompanied by a microwave, sustain those working long hours.

In these days of skill shortages, economic security arises from the skills of the employee, rather than from one particular employer. Because skilled staff have little difficulty in getting jobs which pay enough to cover their needs, an even greater reliance must be placed on means other than financial to attract and retain the best talent.

Social needs depend on the company culture. While not advocating office affairs – they always lead to complications and should be strongly discouraged – many companies encourage informal social interaction outside work, for example with outings, beer nights, company sports teams, and so on. Such events help with informal communications, and tie the company together as a social entity.

‘IT IS WORTH BEING WARY OF USING STATUS-BASED REWARDS’

Status needs are more difficult. Ideally status should be gained from achievement, and from external recognition. However, in the absence of other reinforcement, people will find their own status demarcation, such as minor differences in pay, or trivial things like who gets chocolate biscuits with their coffee, or how close to the front door is their assigned parking place. Some companies deliberately use such indicators as status marks, by, for example, awarding the top parking place to the employee of the month. It is worth being wary of using status-based rewards, however, as they can lead to jealousy and an internally divisive culture, rather than the much preferable culture where status is gained from being part of a world-beating team.

Maslow used the Jungian term self-actualization for the highest level of his hierarchy. Roughly speaking, this means having an awareness and acceptance of oneself, with opportunities to grow and develop, and a belief that tomorrow will be even better than today. In company terms, this means having a career path that continually provides you with new challenges and the opportunity to learn. Many companies have training schemes, and in rapidly expanding companies the opportunity for challenges and career advancement is not usually a problem.

## Management structures

The internal organization and management structure in a company reflect the difference between the two types of management theories, Theory X and Theory Y. Traditional companies tend to use a hierarchical structure, with rigid demarcations between departments, while people-based companies usually have more flexible structures such as the matrix structure illustrated opposite.

‘A HIERARCHICAL STRUCTURE IS CHARACTERIZED BY AUTHORITARIAN SUPERIOR–SUBORDINATE RELATIONSHIPS’

### Hierarchical structure

Hierarchical structures are traditional for Theory X companies. They have the advantage that responsibility and ownership are clear, but hierarchical organizations are typically slow to change, and when they do, such as when the head honcho is replaced, they suffer major periods of disruption.

A hierarchical structure is characterized by authoritarian superior–subordinate relationships, such as in the armed services. Communication is good up and down the tree, although it may suffer from the ‘Chinese whisper syndrome’. However, communication is often poor horizontally, between the branches. Indeed, the typical hierarchical organization only really functions because of informal peer-to-peer horizontal networking – people gossiping around the water-cooler or over coffee.

## Matrix structure

| *Team*<br>*People/skill* | *Home group* | *1* | *2* | *3* | *4* |
|---|---|---|---|---|---|
| Alice | A | L(75%) | | 25% | |
| Bob | B | | L(75%) | | 25% |
| Charlie | B | 50% | 50% | | |
| Dave | A | | 25% | | L(75%) |
| Elizabeth | A | 25% | | | 75% |
| Fred | B | | | L(75%) | |

Percentages indicate percentage of time allocated. L indicates project leader.

A matrix structure is considerably more flexible. A good analogy might be with classes and sets at school. Each pupil has a home class or group for administrative purposes, but is a member of a different set or team for separate subjects, or in our case, projects. Similarly, each employee has a home group for administrative purposes, but divides their time between different teams on various projects.

A matrix management structure allows resources for a new project to be drawn from across the organization, regardless of status. A project leader on one project may act as an adviser on another.

Theory Y companies often use a version of matrix management. However, ongoing responsibility and ownership of resource can sometimes be obscure in such a structure, and should therefore be made explicit, for example by being assigned to a specific home group on a hierarchical basis. Without such assignment of responsibility, things can fall in the gap between one team finishing and another starting.

Matrix-style management structures are good for high-tech start-ups. Start-ups usually begin with a single project, but market pressures and the need not to put all the eggs in one basket quickly lead to the need to support several projects, serviced by the same set of people.

## Groups

At the beginning of a new project or a new company we need to gather a group of people together.

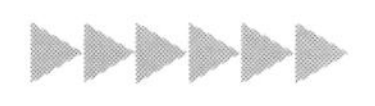

### What is a group?

A group has:

- *Definable membership*. A group is a collection of two or more people identifiable by name or type.
- *Group consciousness*. The members of the group think of themselves as a group, and acknowledge some sort of membership. For example, they might say, 'I'm a member of the Digital Technology Development Group'.
- *Collective perception and identification*. The group has a shared purpose which is the single most defining attribute. There is some common task, goal or interest shared by all members of the group.
- *Interdependence*. Members of the group need to help each other to accomplish the purposes for which they joined the group through interaction and communication with each other, and being responsive and reactive to each other.

Many writers, such as John Adair, Henry Mintzberg, Meredith Belbin and others, have thought about groups, and what makes some more successful than others.

## Work types

A group needs to be composed of many different types. Some people are better at starting things and see the broad picture, whereas others prefer detailed work and finishing a project. A successful group includes elements of both. A definition of key types of people can be found at **www.belbin.com**.

## Developing a team

A group is not a team, at least not at first. A group is still a disparate set of people, and needs to organize itself to get work done as a team.

‘A GROUP IS NOT A TEAM, AT LEAST NOT AT FIRST’

Tuckman (in ‘Developmental sequence in small groups’) models the development from a group to a team in five stages.

**Forming** Forming is the initial stage before individuals see themselves as a team, when the group has first come together. This stage can feel difficult, anxious and unsure.

**Storming** Storming is the stage during which relationships are worked out. It is characterized by challenges to the group leader. Emotionally it is a time of hostility and aggression, with emotions running high.

**Norming** After the storm comes the calm, and the group begins to function as a team, recognizing each person’s strengths and weaknesses. Tasks are organized, and people co-operate, feeling secure and comfortable.

**Performing** The team is now at its most productive. Work surges ahead and people perform well. There is a spirit of openness, mutual trust and support, enthusiasm and inspiration.

**Adjourning** At the end of the project, the work is done, and hand-over processes need to be initiated. It is a bittersweet time, but friendships made continue after the team has disbanded.

## Networking

Teams do not exist in a vacuum. They depend on help and co-operation with other teams to achieve their objectives. For example, a typical development team might depend on an overall architecture team, other development teams whose work interfaces with their own, a customer relationship team to sell and support whatever they are doing, and also finance, legal, building services and other support teams.

There are major pitfalls if teams do not network. These include:

- *Inter-group hostility*. 'We are the Red development team, the hottest on the planet. We never take any notice of what the Blue team says – they must conform to our interfaces.'
- *Insularity and inward thinking*. 'Nothing invented outside of Cambridge can have any value.' 'We have always done it this way. This is the way we do it.'
- *NIH (Not Invented Here) factor*. 'Our solution is the only possible solution.'

TO ENCOURAGE NETWORKING REQUIRES A CONSCIOUS EFFORT, AND NOT JUST A COMPANY NEWS-SHEET

To encourage networking requires a conscious effort, and not just a company news-sheet. Informal social events, such as inter-team bowling or organized pub outings, can help to encourage people to interact and work together. Joint seminars can also be a very good way of enabling your teams to understand each other's work, and will encourage an outward-looking culture.

## Ego-less working

As Weinberg points out in *The Psychology of Computer Programming*, people invest themselves in their work.

The problem is that criticism of the work is then interpreted as criticism of the person, and all sorts of defensive measures happen. For example to avoid criticism one way is not to show the work to anyone. A project manager might well not admit their difficulties until far too late, and the program or project is in real trouble, as they fear to do so would expose their ignorance, or incompetence. Thus minor problems get left until they become major, and cannot be ignored any longer. This effect is also seen in companies running into financial trouble – the directors will sit there in denial, with their heads in the sand until some external agency, such as the taxman, forces bankruptcy.

It is important to foster a culture of trust where it is OK to admit mistakes and ask for help. If people feel unable to admit their difficulties as they occur, small problems can escalate into larger ones, sometimes with very serious effects. It needs conscious effort and leadership to install a culture that can overcome this. People must be made to feel valued and appreciated, and that criticism of their work is not a criticism of themselves.

‘PEOPLE MUST BE MADE TO FEEL THAT CRITICISM OF THEIR WORK IS NOT A CRITICISM OF THEMSELVES’

Regular team meetings can be the perfect opportunity for people not only to catch up on what others are doing, but also to raise any issues and share information. Requests might range from need for information – ‘Hey! I don’t know how to do this bit’ – to requests for resource :‘I know I promised to get this done by Friday, but now I see it will take at least another week, I need another person to help me. Can anyone assist?’ Such meetings can also be a forum for raising more serious issues, and will help to make sure that everyone feels that they can have their say, and are involved. It is essential that such requests are answered, and action taken, otherwise the team members will lose faith in the system.

## Models for teamwork: the surgical team

Creating an organizational structure and defining roles within the group is one of the most difficult stages in forming a team, and it is useful to use models for this procedure. One such model is the Chief Programmer Model, which was documented by Fred Brooks in his book *The Mythical Man-Month* quoting Harlan Mills. This model shows how nine people can be organized to work on a single project.

Although the method was pioneered for programming tasks, it works well for any design process where a single consistent vision is needed. It is a model used by many teams working on complex projects. For a small project, roles can be combined, or people shared between projects.

Brooks uses an analogy with a surgical team, where one surgeon does the actual cutting, or with a flight deck, where one pilot flies the aeroplane. In both cases the rest of the team support the main actor. Note that the pilot and key leader of the group may not be viewed as the head of the group externally.

He proposes the following roles:

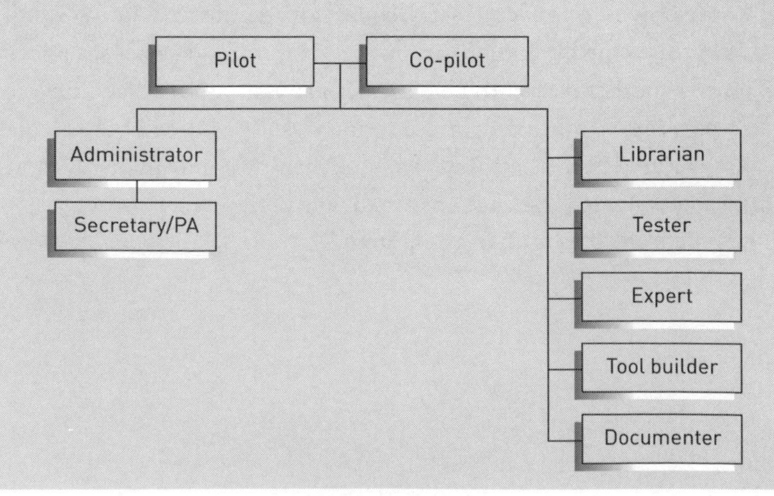

Brooks defines the roles as:

- *The pilot*. The pilot holds the vision, and makes every major decision. Typically the pilot is an experienced and senior person.
- *The co-pilot*. The co-pilot is there to support the pilot and act as a sounding board. They are also an insurance policy in case anything happens to the pilot. The co-pilot is junior to the pilot, typically a pilot in waiting. They may explore alternative solutions and act as the pilot's gofer.
- *The librarian*. The librarian keeps track of the project library, preferably using one of the more modern tools. They ensure the product goes through quality checks and is documented. Also keeps track of all the manifold pieces of paper.
- *The tester.* The tester tests the product and is preferably someone of a sour and mean disposition with a dislike of the pilot, possibly a passed-over pilot candidate. Nothing goes out of the door without the tester's signature.
- *The experts*. There may be more than one expert, but they could be part-time or external consultants, called in as needed. Experts are specialists in particular areas, such as the problem domain, or the target environment.
- *The tool builder*. The tool builder is a position for an apprentice pilot. Although the work the tool builder does may not contribute directly to the final product, it is of great importance. The tool builder helps create the internal scaffolding required for all projects, such as tools, tests, and modifications to the development environment,etc.
- *The documenter*. The rest of the team are techies, and not noted for their ability to write good clear English. A documenter is required to clearly and concisely document the project. Documentation is a vital part of any project, and a speciality of its own.

- *The administrator*. The administrator is the interface between the team and the rest of the organization and the customer. They may even be regarded as the nominal head of the group for administrative purposes. The Administrator gets to wear suits and meet the real customer, and in turn acts as a pseudo-customer for to the rest of the group.
- *The secretary/PA*. The Secretary is notionally assigned to the administrator, but in fact works for the whole group. They also act as office manager and ensure that a good supply of stationery and caffeine is maintained. The secretary is also often the group mediator, smoother of ruffled feathers and culture leader.

‘MANAGERS ARE PEOPLE WHO DO THINGS RIGHT, WHILE LEADERS ARE PEOPLE WHO DO THE RIGHT THING’

(WARREN BENNIS, ON BECOMING A LEADER)

## Managerial roles

Theory Y managers concentrate on leadership and enablement, distributing not only tasks but also the authority required to allow the job to be done. They create an environment where members of their team can take individual responsibility, make their own decisions and get on with the work unimpeded. If anything, the leader and manager is the servant of the team, rather than the other way round, fighting corporate battles for resources and making their views known to the rest of the organization, or to the investors. It's a tough job.

A manager is a leader, providing vision and inspiration and rising above the norm.

‘THE LEADER AND MANAGER IS THE SERVANT OF THE TEAM’

There is no one right way to manage. There are different styles for different people and circumstances. For example, one can draw a continuum between authoritarian and democratic styles.

Authoritarian managers make every decision themselves. They solve their problems alone, and then dictate the decisions. They therefore easily get

## More management theory

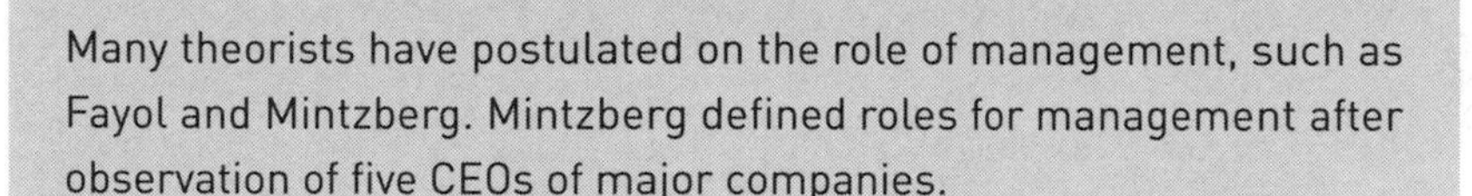

Many theorists have postulated on the role of management, such as Fayol and Mintzberg. Mintzberg defined roles for management after observation of five CEOs of major companies.

He observed how easy it was for managers to get trapped by immediate, apparently urgent activity, getting forced into short-term fire-fighting and 'micro-management'.

If the overall goals are not clearly understood, expressed and shared, the only way a manager can communicate is by directing each small piece.

Micro-managers quickly get overloaded, their staff get demoralized, and their team become inflexible, inefficient and incapable of using their initiative.

It's the difference between setting an auto-pilot and continuously trying to steer the plane yourself.

overloaded, and have a tendency to micro-manage. Their staff get discouraged because their input is ignored, and they have no opportunity to show their initiative.

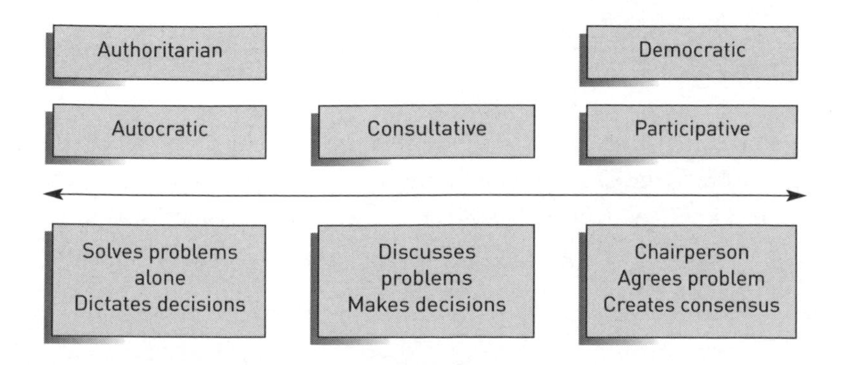

A participative manager, on the other hand, simply acts as a chairperson and referee. A lot of time is spent in meetings and discussions, and decisions tend to be slow and conservative, since the whole group needs to be convinced. Responsibility is diffuse, with no one person owning the decision. This management style works well in voluntary organizations, for example, or for decisions where the staff are considerably more expert than the manager. However, for most commercial decisions there needs to be some single decision maker to take responsibility and act as a champion for the decision.

‘A PARTICIPATIVE MANAGER, ON THE OTHER HAND, SIMPLY ACTS AS A CHAIRPERSON AND REFEREE’

The middle path is for a manager to invite and listen to input, and then to make the actual decision based on the feedback and any other factors. The manager still has to sell the decision to their staff and other interested parties, but staff will feel empowered through their ability to influence the outcome. At the end of the day there is someone accountable, and the decision making is fairly swift, and hopefully consistent.

## Communication

Poor communication is a major cause of most management failures. As soon as there is more than one level of management, games of ‘Chinese whispers’ ensue and messages can become misrepresented or not even communicated at all (an example is given in Chapter 16). Communication failures can of course happen both up and down the management tree, and good communication requires participation from all involved. For instance, senior management may have a clear vision of the strategy, but fail to communicate it, or the people on the ground may know exactly what’s wrong, but are not able to get the message through.

An example of management failure that can be traced directly to poor communication was the *Challenger* space shuttle disaster in 1986. The engineers at Thiokol knew that the seals on the booster would fail at low temperatures, but could not get the message through. In this case, poor communication led to tragedy.

Getting the communication flowing is the single biggest contribution that management can make to the success of an enterprise. This means not only the formal communication, but also things like company newsletters, beer nights, outings, internal websites and notice boards, suggestion boxes, quality circles, collective bargaining, etc. It's also a two-way street. You must listen as well as talk.

Researchers such as Shannon and Weaver **http://sol.brunel.ac.uk/~jarvis/bola/communications/shannon.html** have analyzed communication failure in terms of traditional communication models, examining factors such as noise, filtering, eavesdropping, etc. There is more about the problem of communications in growing companies in Chapter 15.

“GOOD COMMUNICATION REQUIRES PARTICIPATION FROM ALL INVOLVED”

## Meetings

Meetings are mostly a waste of time, and are incredibly expensive. If you add up the day-rates of the people sitting round the table, to say nothing of travel costs and the expense of preparation, you end up with a staggering sum – something like £5,000 per meeting. Do the results really justify the expenditure? Yet people commit to a meeting without thought.

Here are some guidelines:

- Is this meeting really necessary? It's amazing how much can be done by e-mail, or through a couple of phone calls instead. If you must all get together, can you do it by having a teleconference, or by using a videoconferencing tool such as Microsoft's Netmeeting.
- Keep it short. Remember the KISS principle: Keep it Short and Simple. Half an hour might be tolerable, an hour is doubtful, and there are definitely better things to do if it drags on any more. If people won't keep it short, hold the meeting standing up, or just before going-home time, or in the car park on a cold day.

- Organization. Circulate an agenda and papers BEFORE the meeting. No papers, no meeting, unless it's a dire emergency. A meeting is a place for requests and decisions. A meeting is not for briefings, or background information. Do those in writing, or one on one.
- Start on time. Even if not everybody is there. They will be next time. If you don't start on time then it's harder to end on time. Punctuality is the politeness of kings.
- Purpose and structure. The meeting should have a defined purpose and structure, which everyone attending should be aware of. We give some examples below. If you don't know why you are having the meeting, you are lost from the start.
- Minimize interruptions. Some are inevitable, but be sure to have cell-phones turned off.
- Finish on time. People have other meetings to go to. If the business of the meeting cannot be completed in the time allowed, make another time to finish it.
- One page of minutes, circulated the same day. The minutes need only record the actions (by person) and any major decisions, not every word said.

## Brainstorming

One particular type of meeting is the brainstorm. Typically, these are meetings to jointly solve a problem, or work out a new direction. They are usually highly structured in a fashion that ensures that some sort of conclusion will result. However, recent research has questioned the value of a formal approach to brainstorming, claiming the quality of the result is often doubtful.

A typical brainstorming meeting would have the structure:

1. Introductory session, introducing the participants and outlining the problem or issue.

2. Go round the room, with everyone present listing ideas no matter how crazy they may seem. At this stage there is no discussion and certainly no criticism. The idea is to encourage contribution and think 'outside of the box'. Ideas are listed on a whiteboard or slips of paper.
3. Sort the ideas into rough categories.
4. Vote on the ideas. Each person has, say, three votes.
5. Work on the three ideas that came out top. At this point the group might break up into sub-groups, with each looking at either ideas or particular aspects.
6. Re-join for a plenary session, where each sub-group presents its conclusions. Vote for consensus.
7. Write up the conclusions. Include the other ideas as an appendix.

## Formal meetings

Other types of meetings are more formal. In the US, formal meetings are usually governed by some version of Robert's Rules of Order. There is no exact UK equivalent to this structure, although UK parliamentary procedures are governed by Erskine May. You may sometimes hear Chatham House Rules referred to (Chatham House is the home of the Royal Institute of International Affairs) which to all intents and purposes, only constitute one rule, which concerns confidentiality.

When a meeting, or part thereof, is held under the Chatham House Rule, participants are free to use the information received, but neither the identity nor the affiliation of the speakers, nor that of any other participant, may be revealed; nor may the meeting itself be mentioned.

## Board meetings

One type of a formal meeting is a Board meeting. Again, the formal part of the meeting is a place for crispness, for brief summaries (the full report being in the previously circulated papers) and ratification of decisions rather than discussion.

I find the US style of agenda effective:

Call to order

- Attendance
- Minutes of the last meeting
- Matters arising from the minutes

Statutory business (share allocations, formal appointments, etc.)

Reports

- Operations report
- Financial report, including sales
- Business development
- Personnel

Shareholders' issues

Any other business (AOB)

Date of next meeting

## Exercises

**one** *Classic team demo*. Line the class up in groups of ten or so in a double line facing each other, with their right arms and index fingers stretched out. Lay a garden cane across the outstretched fingers. Tell them the task is to stay in contact with the cane, but lower it to the ground. The cane moves up, not down, as each person locally optimizes their contact with the cane. It takes communication for each to move downwards.

**two** *Middle management muddle.* Divide the class into three. One part is the workers, the second are their managers and the last the senior managers. The senior managers can only communicate with the managers, and the managers can only communicate with the workers. All communications must be in writing. Give the workers some materials for a task – sorting cards, for example. Give the senior managers the goal – sort the cards into ascending order. Halfway through change the task – sort into descending order instead. Also, include some snags, for example some cards might be blank, Watch the chaos ensue.

**three** Document the communication paths, both formal and informal in your workplace/group/school

**four** Draw a proposed organization chart for your new business.

## Further reading

Brooks F. (1995) *The Mythical Man-Month*. Addison-Wesley

Drucker P.F. (1999) *Innovation and Entrepreneurship*. Butterworth-Heinemann

Handy, C. (1992) *Understanding Organisations*. Reissued 4th edn Penguin Books

McGregor, D. (1960) *The Human Side of Enterprise*. McGraw Hill.

Townsend, R. (1984) *Further Up the Organisation*. Michael Joseph

Tuckman, B.W. (1965) 'Development sequence in small groups', *Psychological Bulletin*, 63, 384-99

Tyson, S. (2000) *Essentials of HRM*. 4th edn. Butterworth-Heinemann

Weinberg, G.M. (1998) *The Psychology of Computer Programming*. Silver anniversary edn. Dorset House

**http://sol.brunel.ac.uk/~jarvis/bola/culture/index.html** Business Open Learning Archive.

(Maintained by Chris Jarvis at the Brunel University UK. An excellent and useful collection that repays browsing.)

**www.henrymintzberg.com**

(Henry Mintzberg is Professor of Management Studies at McGill University in Montreal, and one of the leading researchers of the theory and practice of management.)

# 10 Managing projects

> 'The best-laid plans o' mice an' men
> Gang aft a-gley,
> An' lea'e us nought but grief an' pain
> For promised joy.'
>
> (Robert Burns, 1759-96, 'To a Mouse')

**SO, YOU'VE GOT YOUR BRAND NEW COMPANY SET UP,** and you are sitting in your not-too-fancy new offices. You've raised some initial cash, and more is promised. You've recruited some bright sparks. Time to relax? Not likely – it is now that the real hard work starts, and you've got to go and do whatever it was you said you were going to.

This chapter discusses:

- the development cycle and various development methodologies
- charts and critical path analysis
- estimation techniques
- monitoring and milestones.

We look at projects particularly from the point of view of tools and techniques evolved to manage software projects, but the lessons are general. Software projects are notorious for being late, and being difficult to manage, perhaps because, in terms of number of decisions, a large piece of software is among the most complex of all entities built today.

“SOFTWARE PROJECTS ARE NOTORIOUS FOR BEING LATE”

## The development cycle

Before plunging in, it is worth reviewing the role of a project manager.

Primarily, a project manager is an enabler, creating an environment that allows other members of the team to work unimpeded. The manager can act as a referee and chairperson for internal discussions, and resolve disputes. They may interpret the vision, but the vision is primarily held by the project architect (or engineer, chief scientist, or whatever the person is called in your discipline). The manager and architect may be the same person, but more usually are not. The project manager may also be a co-ordinator, but more to the point, they create an environment in which the co-ordination can happen naturally. They do, however, interface with the customer and the wider world, and, in larger companies, with the rest of the management and company.

A project manager acts to minimize the risk in the project. Adding a buffer in time and cost estimates reduces risk, but some will still remain. Any project is controlled by three main variables: time, cost and required features, such as reliability. You can optimize any two at the expense of the third, but not all three at once.

❛ FOR EXAMPLE, IF A PRODUCT MUST BE SHIPPED ON A CERTAIN DATE, YOU CAN ADOPT THE 'TRAIN' MODEL ❜

For example, if a product must be shipped on a certain date, you can adopt the 'train' model – the train leaves at a certain time, and those features not on it must wait for the next one. Thus, those features that are reliable enough and ready to ship get shipped at the release date, and everything else waits for the next release. This is a good model where there are frequent releases. Your customers will prefer a solid, reliable release to one that is crammed with features but flaky. The knowledge that an extra feature will be along soon and the guarantee of reliability compensates somewhat for the slippage. On the other hand, if cost is the main criteria, then features, and in some cases development time, must be trimmed to match the resource available. If features are the key issue, don't announce the release date until they are all working reliably.

The diagram opposite plots effort versus time for a typical development project. Of course, real-world projects are often more complex, and made up of a series of components and phases, each with its own development cycle.

At the start, the project is just a gleam in the eye of one or two people. The specification phase works out what is to be built with a description of its

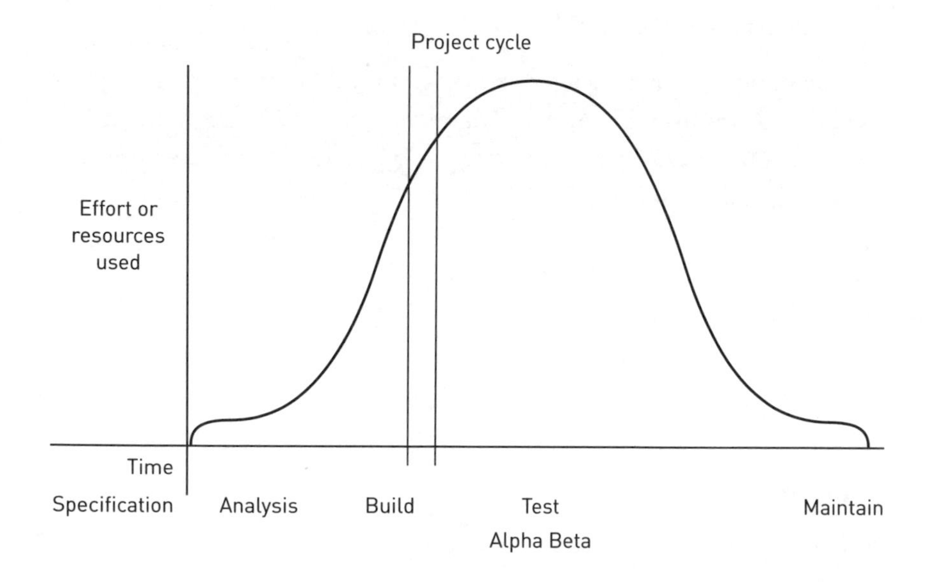


external features and interfaces. The analysis phase figures out the internal architecture and how to build it, with specialists and project managers coming on board. The build phase actually builds the thing, again with more staff, and the project peaks in effort towards the end of the build and at the beginning of testing. Testing is in phases, usually two at least – internal alpha testing and an early release to friends for a beta test. Each test phase will have associated bug fixes, re-works and re-test effort. Finally, the project is finished, and winds down into ongoing care and maintenance.

Think of a building project. At first it is just a gleam in the developer's or architect's eye. The architect produces the drawing, and then the structural engineer, mechanical engineers and others produce the specifications. The builders then build it, and after acceptance hand it to the new owners. There is then ongoing maintenance.

The vertical lines represent the chasm between getting the first prototype working, and engineering a reliable product. We discuss this further in Chapter 11. Crossing this chasm is difficult, as the skills and culture needed on the project change as this boundary is crossed.

## Development methodologies

The project needs to be broken down into manageable chunks, and individual tasks need to be assigned. A traditional approach to this process is to use a top-down (waterfall) decomposition methodology. Some special cases use a bottom-up approach.

### Top down: waterfall decomposition

This approach enables the project to be broken down into smaller and smaller pieces, while defining interfaces between them. For example, if your project consists of designing a particular program, you might start the planning process by breaking the program down into three sections:

> ‘THE PROJECT NEEDS TO BE BROKEN DOWN INTO MANAGEABLE CHUNKS’

- Initialize.
- Do work.
- Clean up.

Each of these is then further broken down:

- Initialize:
  - — Initialize static variables.
  - — Input parameters.
  - — Initialize dynamic variables.
- Do work:
  - — First part.
  - — Second part.
- Clean up:
  - — Output results.
  - — Release resources.

This process can be continued until the components are small enough to program directly.

This example demonstrates sequential decomposition, or step-wise refinement. There are also related techniques, such as data-flow modelling for non-sequential problems.

In the 1980s and early 1990s, a whole new industry and genre of literature was created, devoted to devising methods for reliable problem decomposition. This industry was littered with new techniques such as Yourdan, Martin, Jackson and others, named after their inventors and promoters. Each technique had limited success with certain classes of problems, but none, despite the claims, were universal panaceas; software projects still produced buggy programs and, in some cases, failed altogether. Large projects, and especially software projects, are still notorious for exceeding budgets and timescales, and producing under-performing results.

## Bottom up: increasing complexity

A related approach is the bottom-up technique: instead of breaking the problem down into little pieces, in some cases it is easier to start by specifying the smallest piece, and gradually building up layers of abstraction. Object-oriented design is often done this way. To use the analogy of a building, one way to go about designing it is to first design a standard module (made up out of other standard modules, like standard doors and windows), and then fit the modules into an overall plan.

In practice, real-life designs use a combination of both techniques. Typically, the top-down phases are followed by the bottom-up phases as the various components and interfaces are refined. Sometimes it's most practical to start somewhere in the middle, where the problem is most visible, and work in both directions.

‘IN PRACTICE, REAL-LIFE DESIGNS USE A COMBINATION OF BOTH TECHNIQUES’

## Rapid prototype: successive refinement

The classical methodologies above suffer from a number of problems, a major one being the assumption that the problem you are trying to solve is well defined at the beginning, and that the definition is stable. For a large project, which might take years to complete, the technology will advance and the functionality required will change over the course of the project. With classical waterfall methods, such changes are hard to accommodate and there is no easy method of propagation of changes up the design tree. By the time the detailed specifications have been developed and agreed, they will be out of date, even before whatever they describe has been built.

Projects using waterfall methodologies tend to be all or nothing. There are no natural break-points whereby a reduced functionality prototype can be released. Once started, the project becomes a juggernaut, with the cost of fixing early errors increasing by an order of magnitude at each level of decomposition and development stage. Hence, risk management is difficult for such projects. Projects based on bottom-up methodologies are even harder to control, as it is only at the last minute that it all comes together.

Another difficulty is that much of the design is on paper (or the electronic equivalent). For large systems, such documents are effectively unreadable as the level of required detail obscures the overall intention.

People have now come to realize that building a rapid proof-of-concept prototype, even if it is just a mock-up, has many advantages:

- The users, customers and other stakeholders can see and understand what they are getting and give feedback early in the process.
- Iterating from a prototype is a lot easier than trying to build the finished product directly.
- Risk is more manageable, as there is always a fall-back position.
- Software is rarely written from scratch, but more normally adapted from existing code, so an iterative procedure is natural.

“DEVELOPING A PROJECT ITERATIVELY IS OFTEN THE BEST WAY TO WORK”

Developing a project iteratively, one small step at a time, is often the best way to work. The key is short development cycles, of between six and 12 weeks, which are long enough to make progress, but short enough to

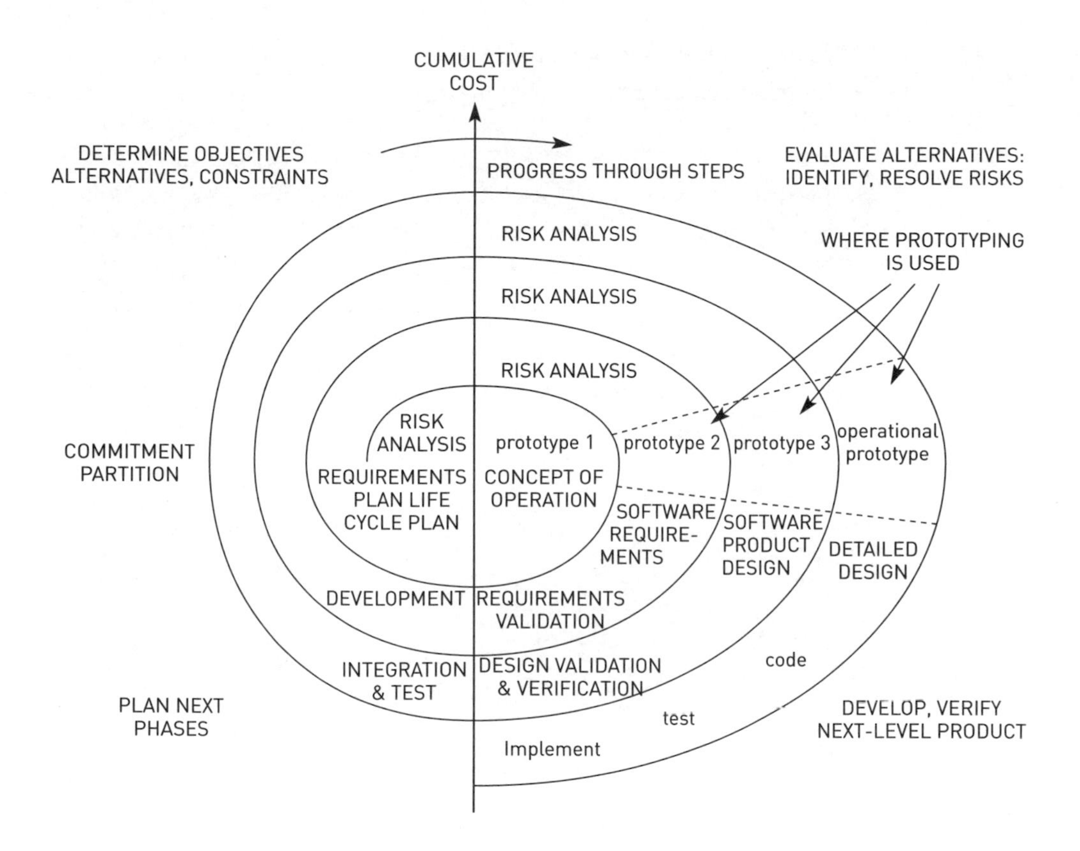


keep focus. The initial prototype can be very simple – a web-page mock-up of the user interface, for example, that is gradually automated, a central routine that works for only one case and with fixed data, or a report outline with sections still to be developed.

The success of the rapid prototype or successive refinement model led in turn to various Rapid Prototyping or Rapid Application Development (RAD)-based methodologies. It also led to the development in the mid-1980s, by Barry Boehm and others, of the 'spiral' model of development.

The original model describes a series of successive iterations, with each adding to the previous prototype. Various later models have refined the basic notion. Most software is today developed with some sort of spiral/successive refinement methodology.

The Microsoft Application Development Process Model (**www.microsoft.com/technet/Analpln/process.asp**) defines four phases in each turn of the spiral, each ending with a well-defined milestone as a review/synchronization point. Effectively each turn is a waterfall development in its own right.

This model defines for each turn of the spiral:

| *Phase* | *Deliverables* |
|---|---|
| Envisioning – agreeing the overall direction and the contents of this phase | Vision/scope document<br>Risk assessment<br>Project structure |
| Planning – design for this phase | Functional specification<br>Risk assessment<br>Project schedule |
| Developing – the actual build | Frozen functional specification<br>Risk management plan<br>Source code and executables<br>Performance support elements<br>Test specification and test cases<br>Master project plan and master project schedule |
| Stabilizing – test, debug, rework | Golden release<br>Release notes<br>Performance support elements |

## Best practice

If you analyze what makes some complex software system projects successful, a number of common threads can be identified. Although these threads are attributable to this specialized project area, the principles apply to any creative development, such as writing a document.

The common lessons seem to be:

- *Multiple cycles*. Don't try and build it all at once. Use several iterations to add features to a stable core.
- *Prototype*. Build a prototype first, so people can see where you are going.
- *Use buffers*. Allow a 20 per cent buffer in both time and resource as things will slip, even in the best controlled projects.
- *Short cycles, focused teams*. Break large projects into several smaller ones. A typical development iteration (once round the spiral) might be six to nine weeks of development, two to four weeks of testing and two weeks of buffer. Milestones should be spaced at weekly intervals, but might be as short as two to three days.
- *Daily builds*. Do a daily build of the software (or equivalent) plus a 'smoke test' – automated test for defects. The typical cycle is for a developer to test their own code, and when they are satisfied with the new feature, check it into the central repository, together with the appropriate tests. These are then included in the daily build. The build is nothing without the associated testing. The term smoke test comes from the early days of electronics, when a new circuit board would be plugged in and the gross defect found by what caught fire. Checking in broken code counts as a sin and is punished, for example by making the perpetrator responsible for the daily builds until a new victim is found.
- *Buddy testers*. Each developer needs a buddy, typically a tester, who tests the code as it is being developed, but is also there to learn, and to bounce ideas off. The tester develops test routines for the feature in parallel with the code.
- *Fixed ship dates*. Having fixed ship dates (although still with variable features) concentrates the mind.
- *Share and enjoy*. Make a fetish of communication. Keep project logs recording why each decision was made. Run project and sub-project

‘HAVING FIXED SHIP DATES CONCENTRATES THE MIND’

## Open source development

Open source development, such as Linux, provide an interesting illustration of modern project management. Here the workforce is transient and disparate, spread across the world. Yet quality software is produced, on a timescale that would be the envy of some commercial organizations. How do they do this?

- Stable core, with incremental additions.
- Common objectives: everyone involved knows what they are doing, and why.
- Good communications.
- Bright, trusted, motivated people.

Interestingly, the way to earn the trust of the community is to do a good job at the mundane work. Every project has its share of dull work to be done – bug fixes, documentation, re-running the tests. Making these things the route to high status ensures that there is never a lack of volunteers for such activity.

bulletin boards. Listen a lot. At the end of each cycle review what went well and what did not, and what could be done differently or better in the next round.

# Tools for planning and tracking

## PERT and Gantt charts

PERT (Project Evaluation and Review Technique) and Gantt charts (named after Henri Gantt) are visual representations of a project. They are essential tools for planning and keeping track of a project, as well as for communicating the state of the project, both to members of the project, and to others.

Of course nowadays, few people draw PERT and Gantt charts by hand, but use one of the many project-planning tools, such as Microsoft Project, which are available for personal computers. If you are not already familiar with such vital tools, you should try and acquaint yourself with one as soon as possible.

To illustrate how such charts can visually represent a project, let's consider the problem of getting up in the morning. You need to leave for work (or lectures) at a fixed time. What is the latest time you can set your alarm for?

First of all you need to write down all the tasks that you must complete before leaving, and the time that they will take (in minutes in this example, although real projects might measure tasks in days). In a big project there might be hundreds or even thousands of tasks, arranged in a hierarchy of sub-projects.

Tasks with zero duration are *milestones*, marking important points, like the beginning and end of the project.

We will start by constructing the PERT chart and looking at the *critical path*.

Draw a box for each task, and join them up with arrows. The task on the tail of the arrow must be completed before the task at the head of the arrow can begin. For example you cannot make the coffee until you have boiled the kettle.

| *Task* | *Duration (mins)* |
|---|---|
| 1 Alarm rings | 0 |
| 2 Wake Up | 3 |
| 3 Get out of bed | 5 |
| 4 Wash | 5 |
| 5 Get dressed | 5 |
| 6 Put kettle on | 2 |
| 7 Wait for kettle to boil | 5 |
| 8 Put toast on | 2 |
| 9 Wait for toast | 3 |
| 10 Make coffee | 3 |
| 11 Butter toast | 2 |
| 12 Eat breakfast | 10 |
| 13 Leave for lectures | 0 |

Tasks immediately before a particular task are called its predecessors. For example 'Make coffee' and 'Butter toast' are predecessors to 'Eat breakfast'. The tasks that immediately follow a task are called its successors. Here, 'Put kettle on' and 'Put toast on' are successors to 'Get dressed'.

The times underneath are the earliest and latest times for the task to start and end.

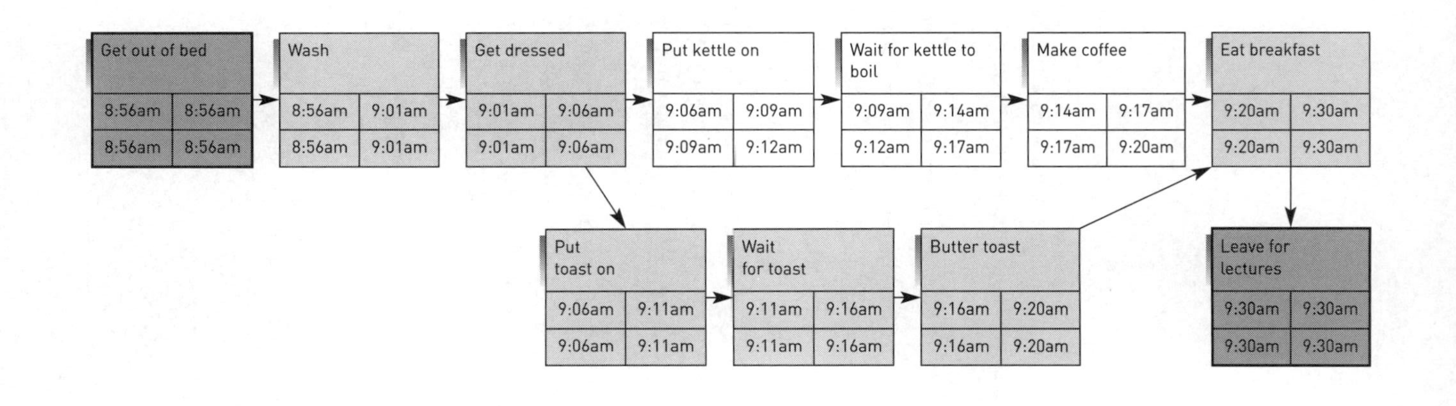

Get out of bed
8:56am 8:56am
8:56am 8:56am
Wash
8:56am 9:01am
8:56am 9:01am
Get dressed
9:01am 9:06am
9:01am 9:06am
Put kettle on
9:06am 9:09am
9:09am 9:12am
Wait for kettle to boil
9:09am 9:14am
9:12am 9:17am
Make coffee
9:14am 9:17am
9:17am 9:20am
Eat breakfast
9:20am 9:30am
9:20am 9:30am
Put toast on
9:06am 9:11am
9:06am 9:11am
Wait for toast
9:11am 9:16am
9:11am 9:16am
Butter toast
9:16am 9:20am
9:16am 9:20am
Leave for lectures
9:30am 9:30am
9:30am 9:30am

| Task Title | |
|---|---|
| Earliest start | Earliest finish |
| Latest start | Latest finish |

You can compute these times by taking a fixed point and working backwards or forwards, subtracting or adding task times as you go. In this case, since the end time is fixed, you need to work backwards, subtracting the task times to give the latest time that the previous task can start. Where there is a fork, as after the 'get dressed' task, take the earlier time. Eventually, you can work back to the first task, which gives the latest start time. You can then work forward computing the earliest start time for each task.

Now you have worked out the earliest and latest start (and finish) times defined for each task. A task cannot start before the earliest time, as its predecessors will not have completed. The task cannot start later than the latest time or the end time will slip. The difference between the earliest start and the latest start of a task is called the *slack*. Tasks with slack can slip within their slack time without affecting the project as a whole. Tasks with zero slack are on the *critical path*, and any slippage will affect the project as a whole. The critical path is shown here in bold.

This could have been made this more complex. For example, the task 'Get dressed' actually represents a sub-project consisting of:

- Put on underpants.
- Put on shirt.
- Put on socks.
- Put on trousers.
- Put on shoes.

Note that it is inadvisable to attempt to put your underpants on after your trousers, or socks after your shoes!

A PERT chart is useful since it shows the structure of the project, and gives some clues as to how things might be optimized, and what the consequences of the tasks slipping are. In this example, it might be possible to wash while waiting for the kettle to boil and the toast to cook, saving a whole ten minutes (see page 210).

Other optimizations are possible. For example, you could introduce additional plant and machinery, such as an automatic tea or coffee-maker, or decide to switch to eating a health bar instead of toast. Maybe you could even eat the health bar while travelling.

A Gantt or bar chart presents the same information but in a different format (see page 211). Here, bars represent the duration of each task, with the slack shown as half-width bars. Milestones are diamonds. Arrows can be added from one task to another to show dependencies, but these can be confusing for large projects. Gantt charts provide a good, quick indication of how the project is progressing, and are more easily drawn on the backs of envelopes than PERT charts are.

Let's take a slightly larger example. A certain software project is planned as three phases, with each phase consisting of analysis, coding and testing tasks. The PERT chart for this project is shown on page 212.

We can associate certain resources with each task, for example, an analyst to do the analysis, and a programmer to do the coding. Either can do the testing. We can now also graph the resources used in each period.

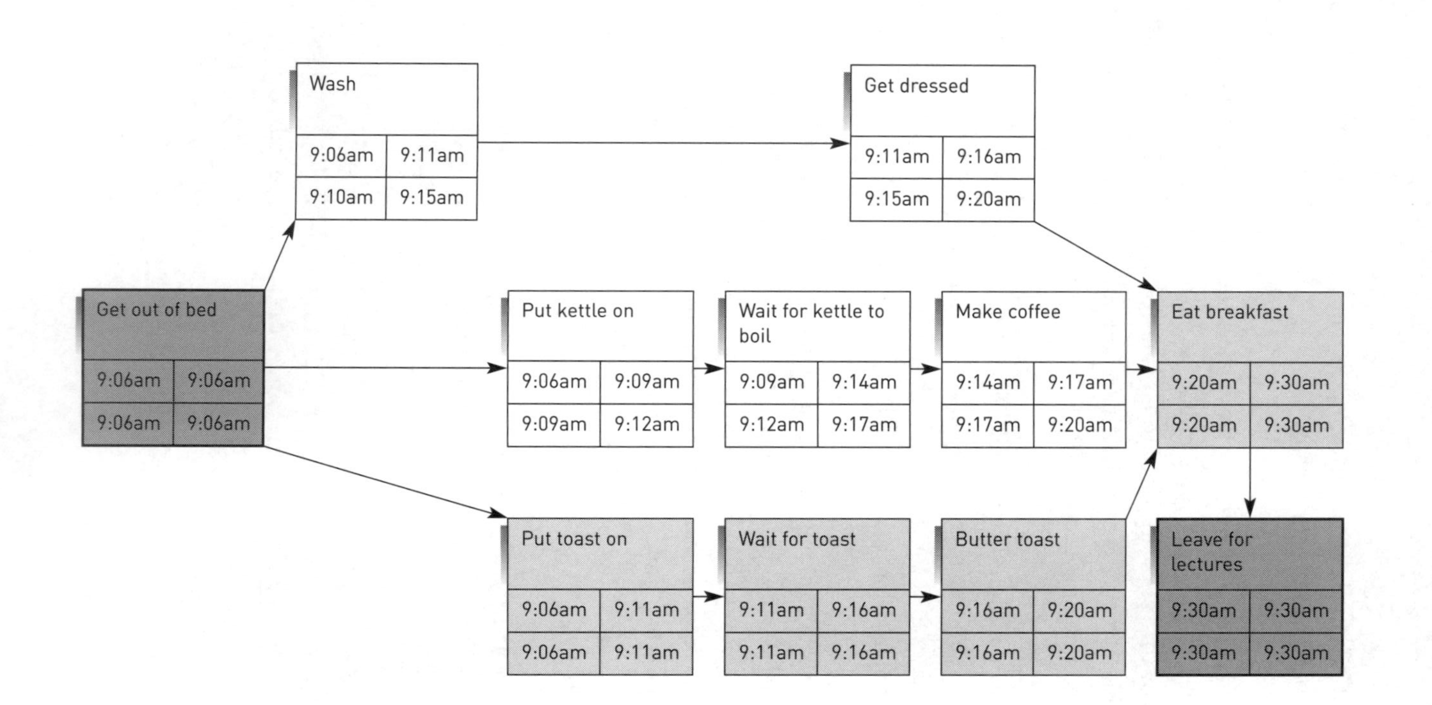

Get out of bed
9:06am
9:06am
9:06am
9:06am
Wash
9:06am
9:11am
9:10am
9:15am
Get dressed
9:11am
9:16am
9:15am
9:20am
Put kettle on
9:06am
9:09am
9:09am
9:12am
Wait for kettle to boil
9:09am
9:14am
9:12am
9:17am
Make coffee
9:14am
9:17am
9:17am
9:20am
Eat breakfast
9:20am
9:30am
9:20am
9:30am
Put toast on
9:06am
9:11am
9:06am
9:11am
Wait for toast
9:11am
9:16am
9:11am
9:16am
Butter toast
9:16am
9:20am
9:16am
9:20am
Leave for lectures
9:30am
9:30am
9:30am
9:30am

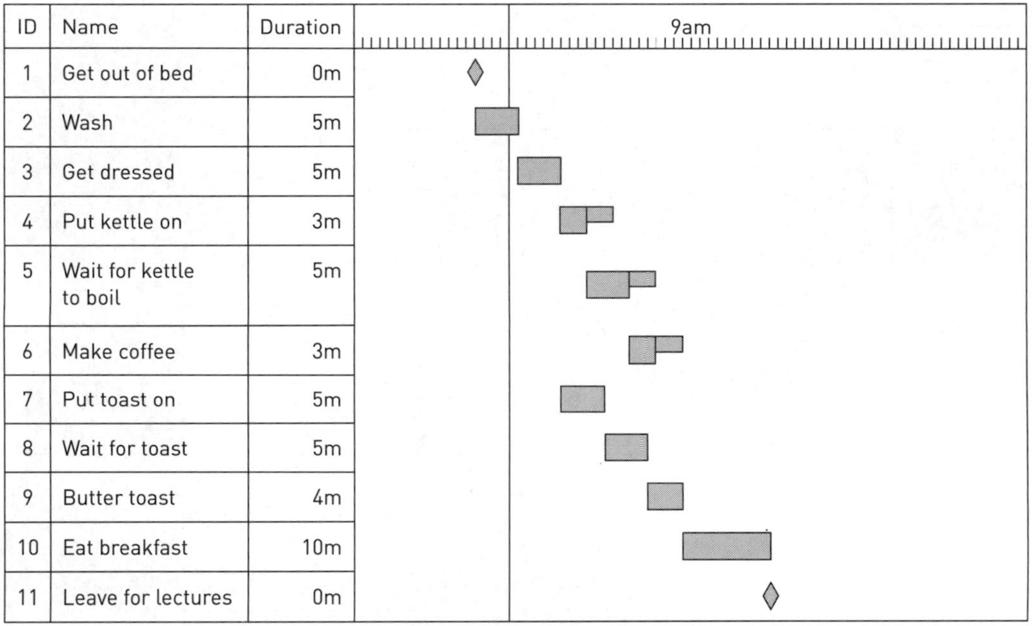

| ID | Name | Duration | 9am |
|---|---|---|---|
| 1 | Get out of bed | 0m | |
| 2 | Wash | 5m | |
| 3 | Get dressed | 5m | |
| 4 | Put kettle on | 3m | |
| 5 | Wait for kettle to boil | 5m | |
| 6 | Make coffee | 3m | |
| 7 | Put toast on | 5m | |
| 8 | Wait for toast | 5m | |
| 9 | Butter toast | 4m | |
| 10 | Eat breakfast | 10m | |
| 11 | Leave for lectures | 0m | |

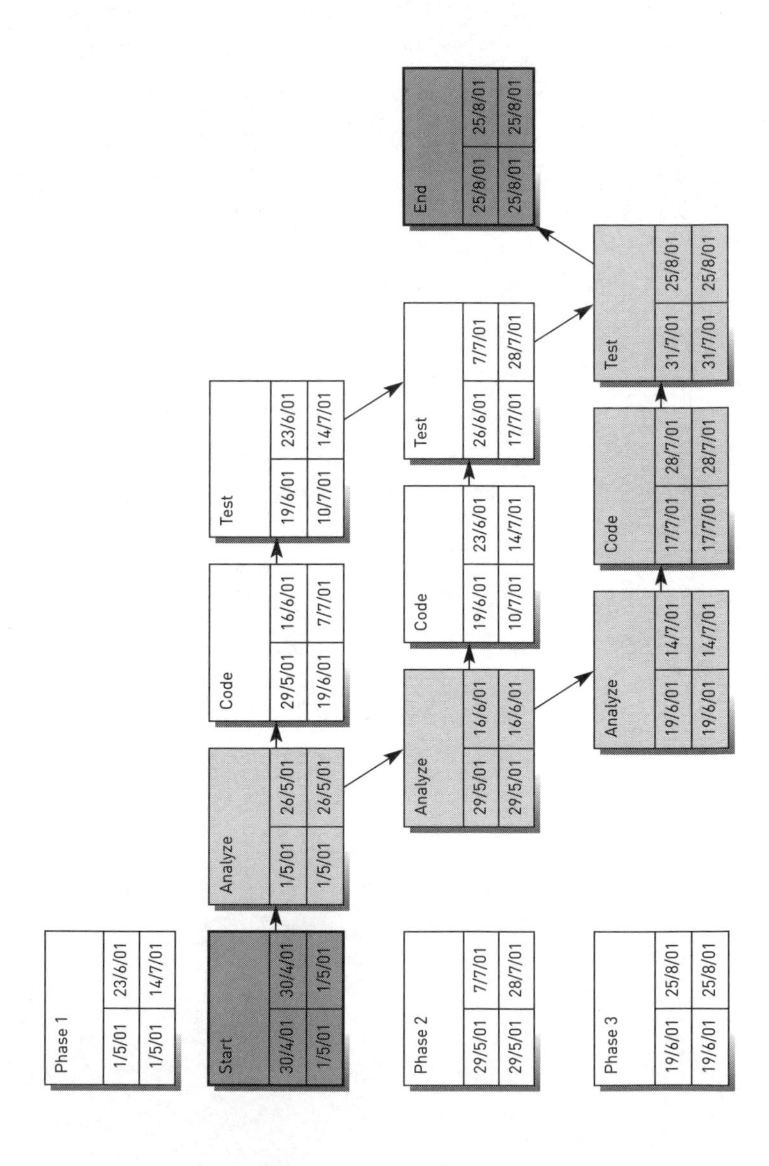
Phase 1
1/5/01
23/6/01
1/5/01
14/7/01
Start
30/4/01
30/4/01
1/5/01
1/5/01
Analyze
1/5/01
26/5/01
1/5/01
26/5/01
Code
29/5/01
16/6/01
19/6/01
7/7/01
Test
19/6/01
23/6/01
10/7/01
14/7/01
Phase 2
29/5/01
7/7/01
29/5/01
28/7/01
Analyze
29/5/01
16/6/01
29/5/01
16/6/01
Code
19/6/01
23/6/01
10/7/01
14/7/01
Test
26/6/01
7/7/01
17/7/01
28/7/01
End
25/8/01
25/8/01
25/8/01
25/8/01
Phase 3
19/6/01
25/8/01
19/6/01
25/8/01
Analyze
19/6/01
14/7/01
19/6/01
14/7/01
Code
17/7/01
28/7/01
17/7/01
28/7/01
Test
31/7/01
25/8/01
31/7/01
25/8/01

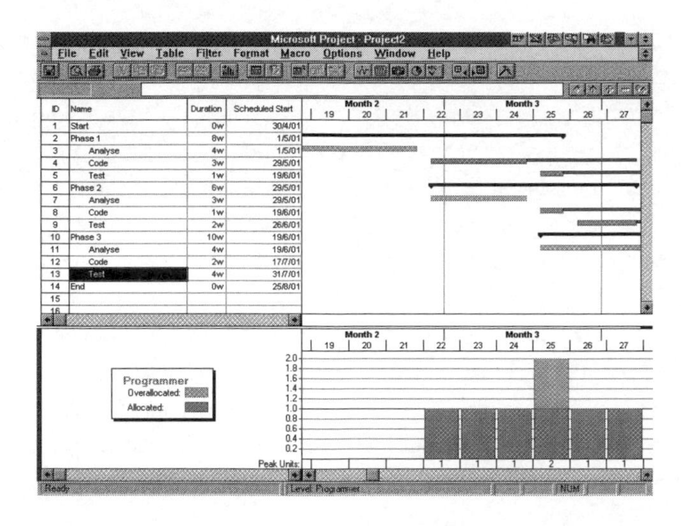


| ID | Name | Duration | Scheduled Start |
|---|---|---|---|
| 1 | Start | 0w | 30/4/01 |
| 2 | Phase 1 | 8w | 1/5/01 |
| 3 | Analyse | 4w | 1/5/01 |
| 4 | Code | 3w | 29/5/01 |
| 5 | Test | 1w | 19/6/01 |
| 6 | Phase 2 | 6w | 29/5/01 |
| 7 | Analyse | 3w | 29/5/01 |
| 8 | Code | 1w | 19/6/01 |
| 9 | Test | 2w | 26/6/01 |
| 10 | Phase 3 | 10w | 19/6/01 |
| 11 | Analyse | 4w | 19/6/01 |
| 12 | Code | 2w | 17/7/01 |
| 13 | Test | 4w | 31/7/01 |
| 14 | End | 0w | 25/8/01 |

However, there is now a problem, as the programmer has been over assigned at Week 25. In order to adjust the schedule to meet the resources available, a process called levelling is used, which has the effect of levelling out the resource utilization graph. There are various strategies that can be employed to achieve this. The first is to slip tasks within their slack to match the resources available. This can be done without necessarily affecting the overall project dates, which is the case in our example.

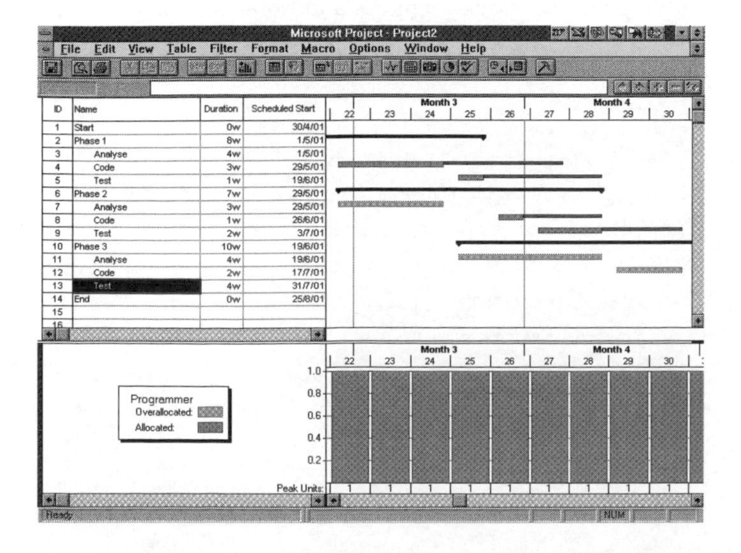


| ID | Name | Duration | Scheduled Start |
|---|---|---|---|
| 1 | Start | 0w | 30/4/01 |
| 2 | Phase 1 | 8w | 1/5/01 |
| 3 | Analyse | 4w | 1/5/01 |
| 4 | Code | 3w | 29/5/01 |
| 5 | Test | 1w | 19/6/01 |
| 6 | Phase 2 | 7w | 29/5/01 |
| 7 | Analyse | 3w | 29/5/01 |
| 8 | Code | 1w | 26/6/01 |
| 9 | Test | 2w | 3/7/01 |
| 10 | Phase 3 | 10w | 19/6/01 |
| 11 | Analyse | 4w | 19/6/01 |
| 12 | Code | 2w | 17/7/01 |
| 13 | Test | 4w | 31/7/01 |
| 14 | End | 0w | 25/8/01 |

However, you are not always so lucky. If the competing tasks are both on the critical path, then either the project as a whole must slip, or extra resources must be drafted in. Both scenarios have disadvantages: slipping the project may have external consequences, and adding resources is never entirely easy. Besides adding cost, the new resources (assuming some are available – good programmers are not easy to recruit) will take time to come up to speed, become familiar with the new environment and learn the task. Also adding new people to a project inevitably slows the existing staff down, as there is another person to train and talk to, and another opinion to take into account, and two people working on the same thing do not work twice as fast as one. Some projects are indivisible, such as the old story of the sultan who wanted a son and heir in one month, and therefore impregnated nine members of his harem.

The phenomenon of *recursive collapse* is more common than one might expect. For example, what does a project manager do, when they realize that their project is running late? Add more resources. However, adding the resources initially slows the project down, and to everyone's consternation, at the next review point, the project is running even more behind schedule. As a result, even more resources are added, and the project slows down even more … In such a situation, you need to consider carefully whether the benefits of adding resources outweigh the additional learning delays and other overheads, although this can be hard to remember when people are screaming for results.

If you must slip a project, do it early, and do it big. With enough early warning, other people – your customers, or the rest of the organization – can adjust their plans. If you leave it until the day before you are due to deliver to tell them that they won't get the product for another three months, then expect trouble. Advertising typically has to be booked six months in advance (Christmas snow scenes are photographed in midsummer), and therefore product launches are planned at least that far ahead. Also cashflow estimates are typically quarterly, which means that you've just blown a hole in the bank balance.

‘IF YOU MUST SLIP A PROJECT, DO IT EARLY, AND DO IT BIG’

Slip big – if you reckon you need another month, ask for three. There is nothing worse than having to go back and ask for yet more time. You essentially get one bite at the cherry. Missing one deadline might be forgivable, as people recognize things can go wrong. However, missing two might not be forgiven, and three means you are looking for another job. If your projects continually miss their estimates you get a bad reputation. Besides, you don't want to go through all that angst too often. Ask for enough time to give you a buffer. Nobody gets shot for delivering early.

Most project packages have built-in automatic levelling software, but these should be used with care. A little real-world knowledge about the project will yield a better result than an automatic system. Maybe tasks can be done in a different order, or estimates revised, or people might even be persuaded to work weekends, just this once, honestly . . .

Most packages allow costs to be associated with a resource, and so project costings can be derived. You can draw up what-ifs, and examine the effect of different strategies on the overall cost. Is it worth buying a bigger machine? The project plan will help tell you.

Project plans such as Gantt charts are not just for the start of the project. They are a constant daily tool. The customer needs another feature? How will that affect the date of delivery? XXX is late again with their delivery (or even more unlikely, early), or module YYY just failed the test and needs re-work. How much does that put other things out?

‘ALL MEMBERS OF THE PROJECT TEAM NEED TO BE AWARE OF THE PROJECT PLAN AND UNDERSTAND THE DEPENDENCIES’

All members of the project team need to be aware of the project plan and understand the dependencies. The latest version should be available to everyone on the project, and everyone should feel that they contribute to it, and have ownership of it.

## Estimation techniques

How do you estimate how long a task will take?

Rule 1: the people doing the work make the estimate. After all, they are the ones who have to deliver. The project manager may then add buffer times and scaling factors, but the source should always be the team doing the work.

Bogus estimates can originate from sources such as:

- Marketing. 'It really has to ship on this date with these features.' The Marketing Department may firmly believe this, but you only have finite resources for the project. Remember, there are only the three variables in a project, of time, resources and product features, and you cannot restrict all three. Something has to give, and this is where good project managers earn their salary and hone their negotiating skills with the marketers.
- Higher management, deciding what the schedule 'should be'. Management can help by making resources available in a timely manner, but they are powerless to dictate the schedule. Trying to do so results in delusion.
- Program management, moving beyond their role of facilitator and remover of obstacles, to inventing something they think will please. If the project plan is not rooted in fact, it is a work of fiction.

A quick test is to ask team members what they think of the schedule. If they are either not aware of it, or scoff, or refuse to discuss it, then the project is in trouble. Other signs of trouble are 'magic dates'. If the schedule miraculously delivers just within the time that marketing or external management demand, be suspicious. Similarly, if several different key components are miraculously scheduled to be completed all on a specific date, be careful. Life is rarely that simple.

## Rules of thumb

Here are some more rules of thumb that people have found useful.

*Software projects: estimate ten × cost and three × time*

Software projects traditionally take three times longer and cost ten times more than the programmer's first estimate. Some people prefer to double the estimate and move to the next unit up. Thus one hour becomes two days, and £100 becomes £2,000. There are all sorts of reasons for this inflation, not least because programmers often only have visibility of the coding part of the development, which is only about ten per cent of the whole thing.

*The 1:3:10 rule*

- 1:cost of prototype
- 3:cost of turning prototype into a product
- 10: cost of sales and marketing

We discuss the process of productization in the next chapter. The consequence of this rule is that new products (and especially start-ups) are dominated by the cost of marketing. Invention is, unfortunately, comparatively cheap. Selling the thing is not.

*Hartree's Law*

- The time to completion of any project, as estimated by the project leader, is a constant (Hartree's constant) regardless of the state of the project.

- A project is 90% complete 90% of the time.

  Often a harassed project manager, might report 'Oh, I reckon we will complete this in about a month', usually without reference to any up-to-date project plan to completion, but as a guess and a hope. This rule illustrates how hard it is to measure something incomplete, and how important is it to make milestone frequent, and concrete, rather than something like 'coding 50 per cent complete.' (Choosing suitable milestones is discussed below.) This rule also demonstrates the need to keep the project plan central, realistic and up to date.

### *The eighty per cent rule*

- Don't plan to use more than 80 per cent of the available resources.

This goes for small as well as big things. We mentioned above the need to allow about 20 per cent in time and resource as a buffer for contingency, and equally only to schedule about 80 per cent of anyone's time, leaving the other 20 per cent for all the other things that impede, but also for marketing support, creative doodling and time to think of the next bright idea. However, this rule applies also to things like office space, disk space, memory capacity, processor cycles and bus bandwidth, etc. Many systems fail if overloaded, and in particular PCs are bad at noticing they are running out of resources. If you do not allow space to breath, you are inviting the dreaded blue screen of death, or at least an unreliable product.

## How long will it take?

How do you tell how long a task will take?

**Experience** By far the most reliable estimate is experience: 'I did one like that last year, and it took 3 weeks.'

The value of experience is a good reason to log both your estimates and the subsequent results, so that you can calibrate how accurate your estimation is. I'm always hopelessly optimistic, usually by a factor of two, so I double the number I first think of.

**Comparison with similar tasks** Even if you haven't done the same task before, you can estimate by comparison with similar tasks, or with industry norms.

Software tasks are notoriously difficult to estimate, as individual programmers can vary by two orders of magnitude in productivity. What takes one programmer an hour, can take another two weeks. Also people tend to estimate just the coding time, not taking into account the design, documentation, testing, meetings and doodling time. The job is not over until you can completely walk away from it, and the paperwork is done.

> 'THE JOB IS NOT OVER UNTIL YOU CAN COMPLETELY WALK AWAY FROM IT, AND THE PAPERWORK IS DONE'

A typical programmer will produce on average 20 lines of fully debugged code a day. For simple tasks, such as a straightforward application, with good tools, this can increase to 200 lines a day. For complex stuff, such as micro-code with a long instruction set, it might be two lines a day, one in the morning and one in the afternoon.

**Decomposition** One problem is that people try and estimate tasks in too big a chunk. If you break the task down into smaller units, and then add up the individual times, even allowing for overlap, you inevitably end up with a bigger number. Smaller tasks are easier to estimate correctly than bigger tasks, and with smaller-scale tasks you are more likely to have come across something similar before on which to base your estimate.

**Plan to throw one away** Earlier in this chapter the value of building a prototype that eventually gets thrown away is laid out. It is always a good idea to allow for at least one iteration in the product. Even if you don't plan for it, there are enough disasters and second thoughts in a project to ensure it will happen anyway. Perhaps you will suffer a disk crash and the backup is unreadable, or there is a fire or a flood, or you discover halfway through that the approach you are taking won't work, or that something vital has been overlooked, or the customer has changed their mind. In all these situations, you will end up doing it again. However, don't despair; the product will be better a second (or third) time around, as by then you will understand the problem.

Fred Brooks, in his book *The Mythical Man-Month*, advises, however, to beware of the second-system effect. It often turns out that there is a problem with the second system a team builds. For the first system out of the door, various corners were cut, features omitted and sacred cows sacrificed to get the product working on time. When it comes to a second version of the product, and there is time and leisure to redo it, there is a terrible tendency to gold-plate the taps. All sorts of unnecessary bells and whistles get added at this stage, and the project will suffer as a result. The Jargon File (**www.tuxedo.org/~esr/jargon/jargon.html**) calls this 'feeping creaturism', where one feature after another gets added, until the whole thing is virtually unusable, or at least not the original clean streamlined design it started out as.

**20 working days per month *but* 200 per year** Project time is funny stuff, and does not add up like normal time. There may be 20 working days in a month, but it is unwise to reckon on more than 200 working days in a year per person. Cynics might say this is because no work gets done in August and in December, but all sorts of thing validly eat into the time available. These include public and individual holidays (did you schedule them?), sickness, training time, conferences, maternity or paternity leave, recruitment of new staff, off-site meetings, meetings which get cancelled, marketing support and just time to doodle.

With 200 days available to work, following the 80% rule (see 'Rules of thumb' box), allow for some buffer time and only schedule 160 actual days.

## Testing

Testing is a religion, and takes as much or more effort than the actual development. It is no longer the case that you can ship some half-finished product to the customer and hope they will test it for you. Nor can you do the development and then do the testing. Testing is a continuous process, and occurs at several levels:

- Unit test, typically by the individual developer before checking in some new feature.
- Daily build and smoke test.
- Alpha test
  - Conformance testing: do the product and the documentation agree? Do they both agree with the specification and the requirements (and the vision behind the requirements)?
  - Stress testing: do they work under load? Do they fail cleanly under abnormal conditions, outside the normal operating envelope?
  - Usability testing: how do people really use it? Can the ergonomics be more effective?
  - Eat your own: use it yourself, for real. Get your family to use it, if appropriate. Is it better than anything else around? If it's that sort of product, is it compelling? Does if feel nice to use? Does it break?
- Beta test
  - Selected external users and early adopters, who are happy to accept a potentially buggy version to get their hands on it sooner, and to join in the testing effort
  - Bug competitions – offer prizes increasing in value over time for bugs found.
- Market test: see Chapter 14.

## Test plan

Every development needs a corresponding test plan. For some industries, such as biotech, the external test plan is pretty well defined by the regulatory regime. For other industries, although there might be good rules of practice, you may need to invent your own test plan. Automated tools are essential, but there will still remain some manual testing. For consistency, there need to be standard scripts, and a standard environment. For software, typical strategies and tools include:

- *Design 'walkthroughs'*. Although tests during the build check whether the design has been implemented correctly, there also need to be processes that check the design.
- *Assertion statements in the code*. Statements in the code that describe the expected state. Some systems can check these automatically at compile-time, others generate code that checks the statement at run-time.
- *Automated test suites*. These can simulate user input and check output.
- *'Monkey tests'*. 'Monkey' tests generate random inputs to the program or sub-system, and check that nothing blows up, that is that no unforeseen errors occur. 'Clever monkey' tests understand more about the program or system, and structure the input, checking the output for consistency as well. By extension, the term can be applied to any system testing with random inputs, for example inputting white noise to an electronic circuit. The term derives from the expression that given enough monkey's typing all the world's literature will be produced. It has nothing to do with animal testing.
- *Coverage checks*. These check that the tests cover every part of the code.

**Test suite** The test suite should cover:

- *Base functionality*. For every statement in the specification there should be a test that checks conformance.
- *Specific bugs*. For each bug reported and fixed, include a test to make sure it has not re-appeared

- *Performance*. Is the speed within acceptable limits?
- *Correct failure*. Does it fail correctly when presented with faulty input?

The test suite should be run each time the code base is updated. Don't just check the new additions – run the whole thing to ensure that no new bugs have been introduced.

Large, poorly designed systems tend to stabilize with a certain number of known bugs, as fixing them may introduce more bugs than they cure. For example, Fred Brooks in *The Mythical Man-Month* reports that the major IBM operating system OS/360 stabilized at about 2,000 known bugs.

LARGE, POORLY DESIGNED SYSTEMS TEND TO STABILIZE WITH A CERTAIN NUMBER OF KNOWN BUGS

## Bug reports and database

You need to get a system in place early to handle bug reports. At the very least you need a report form and a folder, and a process to deal with them. However, with many thousands of bugs reported for a medium sized project, some sort of automated system is preferable, ideally one that integrates with your intranet, and, for example, produces reports as web pages for easy access. Fortunately, there are several such systems available.

The originator of the bug report suggests an importance. A tester on the development team tries to reproduce the bug, and agrees the importance, usually using the following system of levels:

- *Fatal*. The project cannot be released unless this bug is fixed.
- *Important*. Annoying, but not mission critical. A work-round can be used.
- *Useful*. Nice to have, maybe a suggestion for the next release.
- *Not reproducible*. The bug cannot be reproduced, or more information is needed. It should be recorded none the less in case it recurs.

This process is known as triage, even though there are four categories!

## Example bug report form

Date ..............................................................

1. Contact details

Name ..............................................................

E-mail ..............................................................

Phone ..............................................................

2. The problem

Short title ..............................................................

Severity ..............................................................

Sub-system (if known) ..............................................

Build number ..............................................................

Description:

What did you expect to happen?

What is required to reproduce the problem?

---

Date recieved ..............................................................

Entered into database .........................By....................

Acknowledged to user ..............................................

Reproducible? ...................................By ...................

Priority .......................................

Actions:

Reported to developer ............................

Result reported to user .........................

Action closed ..................................

### Action plan for fixes

The next stage is to assign a priority to the bug, and route it to someone to be fixed, preferably the original developer. Senior members of the project, for example the architect and project manager, should meet regularly for a 'bug council'. This might be held daily during some phases of the project, and less frequently as the number of bugs decreases. These people maintain the action plan for fixing all of the high-priority bugs.

When a bug is reported as fixed, the new code version developed and the test for that specific bug satisfied, both the new code and the test for the bug are checked into the code-base. The complete set of tests is then re-run to ensure that the fix hasn't broken anything else. The fix is distributed (or the user politely informed it will be in the next release) and the bug report moved to the fixed pile.

## Recording

Most successful developments centre round similar sets of tools.

Project planning software is discussed above.

### Document and source control system

The central core of any project is the library of documents, design reports and, for software, the source code.

There are various automated version control tools, such as CVS (Concurrent Version System), to administer and keep track of these aspects. Such software will typically ensure that only suitably authorized people can make changes, and will ensure consistency, and that additions have been tested and have the associated documentation, and so on. They will include facilities to enable you to select the latest or any other particular version. Often there are three versions in use at any time: the latest, stable released version, the predecessor version in case backup is needed, and the current experimental build. There may also be individual developers' versions. The software

‘THERE ARE VARIOUS AUTOMATED VERSION CONTROL TOOLS, SUCH AS CVS (CONCURRENT VERSION SYSTEM), TO ADMINISTER AND KEEP TRACK OF THESE ASPECTS’ ensures regular and timely backups, preferably to an off-site location to guard against disaster. For software projects, it is good practice to do a nightly build and smoke test, so that the latest version (the 'daily drop') is available on the team's desktops each morning.

## Logbook

The project needs a central logbook, where all important decisions and events are recorded, preferably with the reasoning behind them. Much of the recording might be by reference or hot-links to external documents. For small projects this might be manual, but for larger ones this should be web based and automated.

## Bulletin board/mailing list

Communication is the biggest problem in any project, and even more so when resources are not physically all in one place, or people work at different times. A bulletin board or several such boards can be very useful, which can be divided into sub-sections such as chat, developer news and corporate news. Mailing lists can also help, but need to be used with care, otherwise the mail system gets overloaded with jokes, requests for lost staplers and other trivia.

## Personal notebooks

What do you carry with you? Some people swear by their Filofax, others by their electronic PDA and many carry their laptops with them wherever they go. These days, many people bring their laptop to meetings, and fancier meeting rooms are equipped with muliple power and internet points for people to plug into, as well as video systems for overhead projection. Although it might be very convenient for people to have online access to the corporate database, their net address books and e-mail documents in a meeting, as well as allowing online minutes to be taken, it can sometimes be unclear as to what they are actually doing on their laptops. Looking round at a the backs of laptop screens is awfully isolating, especially when you suspect that the owners are actually spending the meeting playing solitaire, or catching up with their e-mail.

Personally I use, as do many people with an engineering background, a black and red hardback notebook as a daybook. Anything and everything gets written in it, with a new page for each meeting, dated, with a note of who was there at the top. The back of the book is used for phone numbers and addresses, and the front as a sort of index. Appointments go in a pocket diary, although I can see that I will eventually have to move over to a PDA to record appointments and addresses, if only so that they are co-ordinated with the diary on my desktop which is maintained by my assistant.

## Monitoring

Like budgets, project plans are not something to be done just once at the beginning of a project and then forgotten. They are an essential day-to-day tool.

Monitoring against the plan gives early warning of impending disaster. With early warning you might have time to do something about it, and avoid unpleasant surprises for your bank manager, investors, customers and the other people who depend on you.

Monitoring against the plan needs to be deeply embedded within the culture. It doesn't mean that you cannot innovate, or express originality, but simply that you need to keep the plan up to date with whatever you are doing, and continually consider the effects of your actions. The plan is a key tool for communication both internally and externally, for example explaining to the client just why including this new feature will put the project back three months. Incidentally be honest – the plan should represent your best, even if slightly pessimistic, estimate. Keeping two sets of books, one containing the real plan and another containing a different version that you show to the client, your staff, or your bank manager, will quickly lead to loss of credibility.

“EVERY PROJECT WILL INEVITABLY HAVE ITS SHARE OF BAD NEWS AND THINGS THAT GO WRONG”

To repeat, a key factor is creating a culture in which it is acceptable to air bad news as well as good. Every project will inevitably have its share of bad news and things that go wrong, but trouble shared is trouble halved. The sooner that corrective action can be taken, the smaller the consequences. You need to foster a culture where it is OK to ask for help. Requests must be taken seriously, or they won't be made again. It may be easier for management to stick their heads in the sand and pretend everything is OK, but eventually the problem will become so big that it cannot be ignored. Like the man abseiling down the 100 ft cliff with only 50 ft of rope, reality will get you in the end, and has a nasty habit of not going away. Forewarned is forearmed.

## Milestones

Essential tools for monitoring a project are milestones.

Milestones are definite points along the project's path. They need to be based on concrete objectives and should be independently auditable. An example of a milestone might be 'Module XYZ has passed test PQR', rather than 'Coding 80 per cent complete'. It's amazing how long coding can be stuck at that 80 per cent complete level . . .

Milestones should be set at roughly weekly intervals for each activity. If they are too close together, they take up too much effort, but if they are too far apart, they won't serve to give you the valuable warning of slippage, which they are designed to do.

Beware of fudging milestones, which is a sure sign of a project in trouble. Common fudges are:

- *Moving the target*. 'I know we said August, but I meant the end rather than the beginning of the month.'
- *Premature ejaculation*. 'Well, it almost works. We've only got a bit left to do.'
- *Fudging the objectives*. 'We've shipped the empty case and the invoice, and the product is now checked into the customer's inventory. The customer has returned the product for "repair", and we can complete it at our leisure.'
- *Moving work on* to the next milestone without re-scheduling the project.
- *Excessive secrecy*. 'No, you can't see the test results. Of course we made the milestone.'

Not making a milestone on time should ring big alarm bells. It's all too easy to slip into the attitude of 'don't worry, we'll make it up', or 'we'll fix it if we get the time', but the chances are that you won't. If this is the attitude of the project, be honest, and admit that what will get released on the deadline will be whatever happens to be finished on that date. If this is not good enough, then you had better start rescheduling and finding some more resources straight away! When you get round to planning the next release, you need of course, to set more realistic milestones.

Nobody sets out to deliver a product a year late. It just gets that way, through the accumulation of little slips, and things getting slowly out of control. Ideally, a member of a project team should speak up whenever they are encountering any problems, for instance, if the project is bigger and more complex than originally imagined, and requires re-planning on a more realistic basis, but unfortunately this rarely happens. People's egos are often too great for them to admit that they are wrong, and the project may not be acceptable or profitable on a more realistic timescale. If this is the case it will probably be better to stop now, even if this causes embarrassment. When in a hole, stop digging.

‘HOW DOES A PROJECT GET TO BE A YEAR LATE? DAY BY DAY’

(FRED BROOKS, THE MYTHICAL MAN-MONTH, 1995)

## Review meetings

Project meetings should happen at a regular time each week, or for some projects even daily, sometimes referred to as 'morning prayers'. Review meetings can be arranged in a cascading fashion, from teams, to team leaders, to project leaders, etc. Review meetings provide an opportunity to exchange news, and to ask for help, but they should be brief and to the point. One project manager I know holds his review meetings with everyone standing up.

‘NOBODY SETS OUT TO DELIVER A PRODUCT A YEAR LATE’

# Final word

Despite all your best efforts, and the efforts of the people round you, some things will still go wrong.

Don't despair, it happens to us all. A project typically goes through a number of stages, and being able to recognize these might help.

**Enthusiasm** The new team is full of enthusiasm for the new challenge and the ground- breaking work ahead.

**Disillusionment** The ground-breaking work turns out to be back breaking. Unsolvable problems and insurmountable barriers loom in your path. The vision seems flawed, and the management aren't listening.

**Panic** The critical deadlines loom. People are running around like headless chickens, and doing inappropriate things. Meantime, the programmers work all night for five nights running, only to make some grave error that they were too tired to notice . . .

**Blame of the innocent** The workers are blamed for not delivering. Actually they couldn't have delivered, and they said so, but nobody listened.

**Elevation of the bystander** The kibitzers who stood by and simply watched come out of the mess free of blame, and one of them is then elevated to take charge and sort out the mess.

And so the cycle begins again ...

## Exercise

**one** Go and find a copy of Microsoft Project, or an equivalent project planning tool, and:

- Follow the tutorial.
- Practise by planning how to build a building, for example a garden shed.
- Sketch out what it will take to build your big idea. If the time it will take to get your idea to market, when you plan the detail, is not considerably longer than you originally thought, you have probably made a mistake.

## Further reading

**www.linuxselfhelp.com/howtos/CVS-RCS/CVS-RCS-HOWTO-1.html**
(Tutorial on source code control tools.)

Buckle, J.K. (1977) *Managing Software Projects*. Macdonald and Jane's (Out of print.)

DeGrace, P. and Stahl, L.H. (1990) *Wicked Problems, Righteous Solutions: A Catalogue of Modern Software Engineering Paradigms*. Prentice Hall

Field, M. and Keller, L. (1997) *Project Management*. International Thomson Business Press

O'Connell, F. (1996) *How To Run Successful Projects*. Prentice Hall

11

# From prototype to product

'There is nothing more difficult to carry out, nor more doubtful of success, nor more dangerous to handle, than to initiate a new order of things. For the reformer has enemies in all those who profit by the old order, and only lukewarm defenders in those who would profit by the new order, this lukewarmness arising partly from fear of their adversaries, who have the laws in their favour; and partly from the incredulity of mankind, who do not truly believe in anything new until they have had actual experience of it.'

(Niccolo Machiavelli, *The Prince*, 1532)

CONGRATULATIONS! AFTER MONTHS (OR EVEN YEARS!) of struggle, the thing finally works. You are proudly looking at one of the first prototypes, and very fine it is too. The customers to whom you've shipped the first few hand-built versions also seem reasonably satisfied, although they have a few niggles, and have listed a few extra features that they would like. However, some of your customers have paid, and they are talking about ordering more, so your company at last has income, and hopefully the beginnings of a market.

Time to relax? Nope – you now have three times as much work to do to turn the prototype into a reliable, reproducible product.

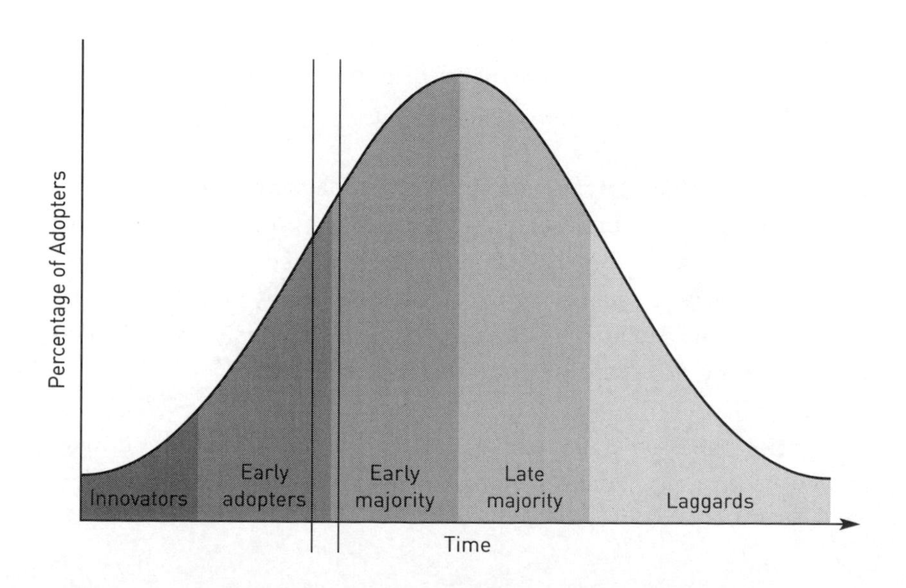

## Crossing the chasm

In Chapter 4 we introduced the idea of the product life cycle. Geoffrey Moore, in his book *Crossing the Chasm*, points out the very real gap that exists between the early adopters and the mainstream users, which is essentially the gap between a prototype and a product. The majority of people wait until a product is popular and has proven reliability before they are willing to risk their hard-earned cash on it.

‘WHEN YOU TURN THE IGNITION IN YOUR CAR, YOU EXPECT IT TO START, PREFERABLY FIRST TIME AND EVERY TIME’

For instance, when you turn the ignition in your car, you expect it to start, preferably first time and every time. You do not care about the details of how the automatic choke and spark enhancer work, or even if they exist – you just want, and expect, the car to start. In the same manner, while a prototype is considered OK by knowledgeable enthusiasts in the ‘Innovator ‘or ‘Early adopter’ categories, it may not seem useable to the majority of people, who will not be prepared to spend time to overcome the difficulties of the early model. While a prototype can be appreciated and used effectively by the developers of the prototype and their friends, it is much less likely to be appreciated by Joe Public. The developers and their friends will have a special knowledge and sympathy with the prototype, and a vested interest. However, Joe Public has a limited knowledge about the product or how it works, and will just want it to work every time. Such customers want to use the product to solve the problem they bought it for, not to help debug it.

Development of a prototype and manufacture of a product involve different skills, and a different mindset. Development of a prototype means doing something different every day. Manufacture of a product means doing the same thing every day, and being able to do it identically every time, getting the details right. High-tech start-up companies typically have difficulty making the transition, or only make it with great pain.

In practice, those early adopters, your first customers, can be your worst enemy. Because they are likely to be techies, they will demand new fancy features, which could divert your effort away from making the underlying product solid and reliable, which are the aspects that will make it sell. The majority market doesn't need the fancy features, and will be happy with the boring basic model. They also want the product to work every time, and will require it to do whatever they bought it for, and certainly do not want to troubleshoot it. For a fledgling high-tech company, it can be very difficult to resist the siren voices of the early adopters offering actual cash to innovate new features. After all, innovating new features is the core talent of the company, rather than using the company's resources to consolidate and finish the existing product for the majority market. However, it's the last ten per cent of that consolidation that takes 90 per cent of the effort.

These considerations apply to companies selling a service just as much as to those with a physical product. Consider a bank or a share broker for example. There is a world of difference between a bespoke service, with a few customers receiving individual attention, and a mass-market high-street standardized service, with set routines.

“THESE CONSIDERATIONS APPLY TO COMPANIES SELLING A SERVICE JUST AS MUCH AS TO THOSE WITH A PHYSICAL PRODUCT”

## Whole product

One of the key differences between a prototype and a product is embodied in the notion of a *whole product*. A product is more than just the actual box you ship. For instance, when you buy a car, you expect that you will be able to service it easily, spare parts will be available, fuel will be easy to buy , and accessories will be available to enable you to customize it. This is no different from anything else. Along with your product you also need to set up whatever it takes to give a good all-round service. This might involve anything from in-store support, helpdesks and service agents, to the marketing of spare parts, running of training courses and fan clubs, or production of magazines. The product needs to exceed customer expectations, not just meet them. It needs to work first time, every time, straight out of the box. It needs to be easy to install, set up and configure, and should come complete, figuratively speaking, batteries included. There should be no nasty surprises, like blue screens of death or unexplained errors with no hint of corrective action. For some things, like games, this means having regular upgrades available, and a support community. For other products, such as cars or investments, this means having an active after-market.

The complete product is everything that the product might one day become, and all the ancillary and support services that go with it. These can range from the trivial, such as the T-shirt, to the not so trivial, like add-on kits and consumables.

The Visio drawing package is a good example of how an original product can develop into a range of products. The Visio package is, basically, a competent stencil-based 2-D drawing package. However, because it was designed to be extensible, a number of third parties have developed specialized stencil sets from the original product, for example 3-D kitchen layout stencils, or project planning tools.

It is often the case that product developments such as these are unexpected, and happen by chance. For example the Quake engine, originally developed for the well-known game, is now used as a tool for visualization of architectural drawings.

## The role of customer acceptance: ACCTO

Everett Rogers, in his ground-breaking book *The Diffusion of Innovation* looked at over 5,000 case studies on the introduction of new products. He observed that 80 per cent of new product failures are due to lack of customer acceptance. He developed the ACCTO criteria for customer acceptance as a result. Here is a summary.

**A – Relative Advantage** Relative advantage is the degree to which an idea or a product is perceived to be superior to that which it replaces. The advantage need not be financial, but may confer social prestige or increase your sense of self-value, or offer some other benefit. For example, Rogers noticed that there was one make of grain silo that was considered more expensive and prestigious than any other make. In fact, this was so much so that some farmers in the US Midwest would purchase two silos, even though they needed only one.

Marketing can boost the perception of a relative advantage. Rogers found that relative advantage was the single most important factor in the spread of an innovation, with all other measures simply reducing the uncertainty of the perception of the advantages. This concept also explains why preventative measures, such as safe sex, are hard to sell, since the perception of advantage is both delayed and diffuse.

Relative advantage of a product or innovation can also be enhanced by incentives, either to the purchasers or to the sales force or change agents.

**C – Complexity** Can I understand it? Complexity is the degree to which an innovation is perceived as difficult to understand or use. Early personal computers suffered from this perception, and breaking down this barrier was one of the reasons for the success of the Apple Macintosh. The customer needs to be able to understand how to use what they are purchasing, and what advantages it has for them. If the customer perceives that the thing is too difficult for them to use, then they won't buy it.

**C – Compatibility** Compatibility is the degree to which the innovation is perceived as being consistent with the needs, values and experiences of the adopters. The product needs to be compatible with the user's environment and beliefs. For example, 240v UK standard electrical equipment has little market potential in the US where 110v is the norm. Similarly, new software needs to run on existing computers and read existing file formats, or otherwise integrate with existing data. New ideas can only be explained in terms of existing understanding, which limits their rate of adoption. However change, when it comes, happens very fast – for example when Eastern block countries were allowed access to Western technology. The new acquirer wants the very latest and fastest gismo, skipping the intermediate steps. Therefore, there is no market for old 386 computers in developing countries.

‘NEW IDEAS CAN ONLY BE EXPLAINED IN TERMS OF EXISTING UNDERSTANDING’

**T – Trialability** Trialability is the degree to which the new innovation may be experimented with before commitment. The customer needs to be able to test the product without risk before purchase. Hence, free trial periods, 90-day no quibble money-back guarantees, test drives and a host of other mechanisms to achieve this. For distance selling, such as over the web, there are particular difficulties.

**O – Observability** Observability is the degree to which the results of the innovation are visible to others. The customer needs to have visibility of the product, and of the resulting benefits. Advertising and customer education can help explain what they should look for. If, having tried it, the benefits, real or imagined, are not obvious to the customer, they are likely to reject the innovation.

## Software as service

Current thinking deems that mass-market software provision is a service, instead of a high-volume, low-margin manufacturing activity. Instead of buying the upgrade to your operating system or wordprocessor every two years for £99 in addition to any service agreement that you might have, it might be better to rent the system for say, £50 a year. This model allows for the transition of service provision, as you upgrade your systems and get new facilities. For example web and mail servers are often provided on a rental basis by the ISP, rather than locally. Provision of substantial functionality in the network via ASP functions, rather than locally, is becoming increasingly common.

> ‘THE FREE SOFTWARE MOVEMENT HAS LONG RECOGNIZED THAT A VIABLE BUSINESS MODEL IS TO GIVE THE SOFTWARE AWAY (WHICH SATISFIES THE TRIALABILITY CRITERION ABOVE), AND TO MAKE YOUR REVENUE ON SERVICE AND SUPPORT ELEMENTS’

The free software movement has long recognized that a viable business model is to give the software away (which satisfies the trialability criterion above), and to make your revenue on service and support elements. With downloading over the net as the means of distribution, the marginal cost of manufacture asymptotes to zero, and so, inevitably, will the price, leaving only the service elements as sources of revenue.

## Productization

Productization is the process of turning the prototype into a real product – one that can be reliably manufactured to be the same each time, and that will work every time, everywhere. It's usually reckoned that this process takes about three times as much effort as building the prototype in the first place.

Here are some of the things that need to be done.

### Generalization

Developers typically test the product in optimum surroundings. They have fancy computers with nice big high-resolution screens. However out in the field, conditions are often far from ideal, with a lot of ancient tin still in use. Testing out all the possible combinations that might be found in the field is difficult. Building a lab even with just the major variants is expensive and time-consuming. To some extent, this is what the beta-testing phase is for, but even before it gets that far, the product should be tested under the majority of the conditions it will meet, and in particular under marginal conditions and under a configuration of minimum support. Does the product work on all target systems? What hardware variants/constraints does it support or require? What operating system variants will it work with?

For example, do your web pages show up well on low-resolution displays? How about on WebTV boxes? (WebTV publish useful guidelines and simulators; for a discussion on what makes an acceptable web page see below.)

For software developed for PCs, does it work on all flavours of Windows (95, 98, Me, NT, 2000, and XP ...)? Will it work with emulators like Citrix? Is it fussy about configuration and additional devices? Is the installer robust? What happens if you try to install it over an existing installation? Is there an uninstall option? What happens when the PC is short on some resource – memory, disk, stack ?

For Unix systems, the situation is still more confused. Although Linux is widely used, especially in the Red Hat variant, there are several other flavours out there, besides other Unix-based systems such as Sun's Solaris, HP's HP-UX, IBM's AIX and the various BSD clones and other variants. There are several flavours of graphical user interface (GUI) of Linux, with the competing camps engaged in wars with each other, carried out with religious fervour. This lack of uniformity is the major factor holding Unix back from world domination.

**Internationalization** For products or services that are to be sold in more than one country, you need to go through the process of internationalization. Although modern programming practice makes this easier by making the user interface soft, with most of the text and messages in a resource file, there is still considerable effort needed in translation. It is also a lot of effort to ensure that the translation is complete, and that, for example, no error messages have been accidentally bound into the code, or are lurking in some third-party library sub-routine. Not only does the product need to be checked, but all the documentation, packaging and labelling needs translation as well. This can be a large job, as the screen-shot illustrations in the manual may need to be re-done. You should try and get a native speaker to check translations, as subtle cultural differences can have unfortunate consequences. There are many examples of this given at **www.intuitive.com/taylor/gs/gs-chap5.html**, such as the unfortunate Japanese soft drink called Sweat, or car names such as Nova, meaning 'doesn't go' in Spanish, or the Fiat Uno in Finland, where Uno means garbage. Even between English-speaking countries there are sources of confusion. For example Durex is well known as a contraceptive in the UK, but is known as adhesive tape in Australia.

❛ YOU SHOULD TRY AND GET A NATIVE SPEAKER TO CHECK TRANSLATIONS, AS SUBTLE CULTURAL DIFFERENCES CAN HAVE UNFORTUNATE CONSEQUENCES ❜

Languages with non-roman alphabets have their own problems, as do graphic icons. The Apple trashcan icon caused confusion as it looks like a UK post box. Similarly, mail icons with flags, such as the mail system in SunView, were misunderstood in many countries, as, unlike US post boxes, the UK postal system does not use flags, and in other countries the flag raised indicates mail for collection by the postal system, rather than delivery.

You also need to ensure that you check out the correct number, currency data and time formats for different countries. This can require a certain degree of subtlety, for example, in making sure that enough space has been allowed for small denomination currencies like the lira and the yen, and that you have allowed for details such as currencies where the currency mark is at the right rather than the left of the number.

There may be alterations needed due to gross feature changes, for instance, putting the steering wheel on the other side of a car may mean that some controls become unreachable, and therefore need moving. Different countries have different telephone, power plug and voltage standards, and even TV formats vary. Other variables include safety requirements – especially for electrical and fire protection, standards of allowable radio frequency emissions, packaging requirements – which in some cases are contradictory, and even things like the contact positions of smartcards, which are different in France.

The whole process of adaptation to local standards is known as *homologization*, and is inevitably tedious, expensive and time-consuming. There are surprisingly few internationally recognized standards or test authorities, even within apparently free trade areas such as the EU. You may need your product to be re-tested for safety, emission and telecom standards in each country, and sometimes each state, in which you market. There are other sensitivities from religious or ethnic backgrounds, for example crescent-shaped buttons can cause offence in Muslim countries. Standards of allowable clothing, adult content, language, political opinion, and sale of products such as alcohol, drugs – both recreational and prescription – tobacco and explosives, including fireworks, vary from place to place. This also applies for services, such as financial, legal or medical services, where local certification is needed. For internet-based services, these are particular problems, and positive action may be needed to exclude users from prohibited places.

## Documentation

We discuss documentation in the next chapter. Getting the product finished and checked against the documentation is one of the big hassles of turning a prototype into a product. The standard of documentation needed for a product is much higher than that needed for a prototype. The documentation may need to educate new users to the wonders of the product.

## Legals

There is a whole lot of legal work associated with getting a product out of the door, in various different areas.

**IPR generated** Patent activity can take up a lot of time, and must be done before the product is 'published'. However, that is not the only form of IPR that needs work. Are all the copyright notices in place? Trade and service marks may need to be registered in every jurisdiction in which the product is to be sold. Advertising, flyers, point-of-sale material, manuals and web pages should be checked by legal advisers to ensure that only valid claims are made. Pay particular attention to what is available to a customer before they purchase as they may rely on such material to help them make their decision, and can sue if the claims made turn out to be untrue. This could include what is written on the box, as well as what you put in your advertising.

**IPR used** Any IPR that is used needs to have a licence on file, and the product should be checked to make sure that it does not infringe the licence conditions. You also need to check that you are not inadvertently using somebody else's existing IPR, by doing a patent search for example.

‘ANOTHER THING TO MAKE SURE OF IS THAT YOU HAVE LICENSED, AND ARE COMPLYING WITH THE TERMS FOR ANY SOFTWARE LIBRARIES THAT YOUR CODE USES’

Another thing to make sure of is that you have licensed, and are complying with the terms for any software libraries that your code uses. Even public domain software may have conditions attached. For example the Free Software Foundation's GPL imposes strict conditions when the software, such as Linux, is incorporated into a product. Many software tool vendors, including Microsoft, impose conditions on the code produced using their tools. Pictures used from a picture library or copied off the web are likely to have conditions of use attached, and fall outside the fair use provisions of copyright.

**Licences, contracts, liability** Your legal team needs to originate, or at least approve, the various licences and contracts for the sale of the product, which might include an end-user licence agreement (EULA), and licences to distribute the product.

There will be lots of contracts with regards to supply and manufacturing, as well as advertising and shipping. When you move into volume manufacture, big numbers can be involved. Insurance for product liability and membership of the appropriate trade bodies will be needed, and you may require some statutory approvals as well. There may be health and safety or environmental approvals needed too.

It is worth repeating that liability for death or injury cannot be disclaimed. For software or hardware that might be life critical, aeronautical or medical applications for example, special care is needed.

## Packaging

You need to get the packaging designed and manufactured. Surprisingly, this is often on a long lead-time compared to the contents. The artwork needs to be developed, and this may involve pictures of the product – or at least of the prototype, since the product probably doesn't exist at the point when the artwork is designed.

For some products, the pack has the important task of selling the product off the shelf, and catching the eye of the passer-by. You need to decide what wording is on the outside of the box to tell the potential customer what they are getting before they open the shrink-wrap. It also needs to be cleared with your legal advisers. Some products may need anti-tamper or other special packaging.

You need to decide what is going to go into the box. Besides the main component what collateral does there need to be? This could include a manual, a quick-start card, cables, adapters, plugs, a warranty card, a welcome video or CD, etc. Will it all fit in? Is there room for last-minute changes or correction sheets?

## Manufacture

Whole books have been written on manufacturing set-up. Most high-tech companies will sub-contract the actual manufacture to specialist companies. This does not remove the problem, however. Sub-contract manufacturers are cookie cutters, and are not paid to think. If you do not specify the product correctly, it will come back wrong. I remember a batch of circuit boards that came back perfect, but mirror image, as I had not specified which side was top. That said, there are many excellent sub-contractors, with whom working is a pleasure.

Mass manufacture is a difficult art. Making lots of anything is difficult, and there is much scope for things to go wrong. If you plan to sell 100K units of your product a year, that is roughly one a minute for every minute of the working day. Look around you. Can your facilities, actual or planned, cope with that number? Can you physically shift and sell that number? Will you be able to ramp up fast enough? Have you got enough warehousing?

“SUB-CONTRACT MANUFACTURERS ARE COOKIE CUTTERS, AND ARE NOT PAID TO THINK”

Software will need the production of CDs, manuals, and possibly videos, together with the packaging. Almost universally, such accessories are sub-contracted to specialist printers for runs of any volume.

There are specialist fulfilment houses which offer warehousing and distribution services. Such houses will take in stock from the manufacturer and orders from the web, or from a call centre, and send out the appropriate goods to the customers. They may even handle the credit-card payment as well, although this will be for a fee. Indeed, it is possible to run a completely virtual manufacturing and fulfilment operation, with everything provided by sub-contractors. This has the advantage of flexibility and reduced capital need, but in turn may be higher in running costs, as the sub-contractor's profit has to come from somewhere.

Electronics are often manufactured off-shore in a country with low labour costs, such as in the Far East. In practice, the labour cost difference is not now that high, but such countries have now developed manufacturing expertise that allows extremely competitive quotes, and are often assisted by generous government grants and export schemes. For example, in Taiwan, long credit periods are sometimes offered, with the sub-contractor effectively taking the stocking and tooling costs. In turn, the Taiwanese company might contract the actual manufacture to a factory that may even be in a different country, such as the Guandong province in China.

Off-shore manufacture is not without its problems. One of the greatest difficulties is that of communication, as the people you need to deal with may not be able to speak your language with any fluency, and culture values may cause problems. It might be wise to do the initial manufacture closer to home, in a place which will be easier to visit and control, at least until you have got the bugs out. This may be more expensive, but it is certainly cheaper than having container-loads of the wrong thing delivered.

The cheapest country is not always the best: the local infrastructure and labour market may not be able support your needs, or the country may be unstable politically. For example, Indonesia is regarded as unsatisfactory, but Vietnam is considered as a good place to manufacture, especially as it has been infected by the love of the dollar, and many people speak English. The governmental trade agency (in the UK the DTI) can advise on such matters, and the host government's embassy can put you in touch with reliable partners. Don't be shy about using them: they are there to help.

When manufacturing in a distant location, you need to take shipping times into account. Sea shipment can be up to eight weeks from the Far East. Air shipment is more like a week, but much more expensive. One strategy is to ship an initial batch by air, to start sales rolling, and the rest by sea, given that the sales ramp-up will not be instantaneous.

“WHEN MANUFACTURING IN A DISTANT LOCATION, YOU NEED TO TAKE SHIPPING TIMES INTO ACCOUNT”

**Testing** A major part of the organization of a manufacturing process is setting up suitable automatic test procedures. Quality control issues are discussed below, but manufacturing quality control is quite different to design quality. Of course, good design and design for manufacture help, but making the same thing right every time is a difficult discipline, and an art form of its own. Although the manufacturing sub-contractor can help, at the end of the day you as the customer have to take responsibility and get involved.

## Marketing materials

Generating the marketing collateral takes effort, and maybe the employment of specialists like advertising agencies and PR firms. We look at marketing in detail in Chapter 13. You will need leaflets, brochures, advertisements and web pages as a minimum.

For each of these you will have to generate the text and the artwork, and then get the proofs approved by all the stakeholders. This process can take considerably longer than seems reasonable. Everybody will have a strongly held opinion on the exact shade of background colour needed, or the layout of the buttons on the web page.

You may need legal checks – for example are all the statements true, and are all the copyright or trademark marks in the right place? For some products, such as financial or medical products, you may need regulatory approval as well.

If you are selling off the page (or even off the web page), you will need to set up the mechanism required for doing this, and even if you sub-contract this function to a fulfilment house, the organization of this will still take a considerable amount of effort. For instance, you will need to write the scripts for the call centre people, and at least train the trainers about the product. Setting up your own call and fulfilment centre, however, is only recommended for masochists. Other things that you may need to sort out include the facility to accept credit cards and bank direct debits, for which you will need to ensure that you comply with all of the various regulations (such as the EU Distance-Selling Directive) and the more stringent requirements of the credit-card company itself. You may need to set up secure audit trails and other fraud prevention for this purpose.

## Shipping

There are some technical terms connected with shipping that you might come across. Common ones are:

- *FoB*, meaning Freight on Board. The manufacturer pays for delivery to the ship, aircraft or whatever, and thereafter the goods are your responsibility. You have to pay shipping, insurance, freight handling on arrival, duty, etc.
- *DPD*, meaning Duty Paid Delivered. The manufacturer takes responsibility all the way up to your front door.
- *Waybill* (or for air shipment an Air Waybill or AWB). A document evidencing the contract for transport of the cargo. It can also act as a receipt for the goods. For example, a FedEx receipt doubles as an AWB.
- *Bill of lading*. This can be three things:
  - A receipt for goods, signed by a duly authorized person on behalf of the carriers.
  - A document of title to the goods described therein.
  - Evidence of the terms and conditions of carriage agreed upon between the two parties.
- *Letter of Credit* (LoC). Instructions to a bank to pay money on sight. LoCs can be unconditional, or have certain conditions attached to them, such as only to be paid after a certain date or event, or if additional documentation is present.

These things were worked out in the days of sailing ships, when messages took a long time to arrive, yet trade and the banking systems still had to continue to operate in the meantime.

A typical sequence of events for FoB trade might be:

- The customer places an order with the manufacturer, accompanying it with an irrevocable letter of credit, payable, say, 90 days after the date of the relevant bill of lading. This guarantees that the manufacturer will be paid, assuming that they deliver the goods to the ship. The bill has been issued by the customer's home bank, drawn on a correspondent bank near the manufacturer. Naturally, the bank will insist on either a charge to cover the risk, or a deposit for the amount of the bill, or being bankers, both.
- On delivery of the goods to the ship, the carrier issues a bill of lading. The manufacture presents the bill of lading and the LoC to the bank, (the manufacturer's correspondent bank), who will pay the manufacturer at the time specified, and then reclaim the money from the home bank.
- The bill of lading is sent to the customer, who uses it to claim the goods from the carrier.

Of course, the customer is still at risk, as the manufacturer could have provided the wrong or shoddy goods, and therefore you may want to arrange an inspection before the issue of the bill of lading by a local agent.

Many electronic formats of equivalent documents exist in the banking systems of the world (such as **www.bolero.net**), and every now and then, various people, for example David Chaum, (**www.chaum.com**) propose schemes for unforgeable bearer documents for trade on the internet, especially for low-value trades, such as postage stamp equivalents. To date, none have achieved critical mass, but this area remains a fascinating research topic.

The order-taking and fulfilment mechanism will also need to be tested, and the effectiveness of your advertising and other marketing collateral can be tried out through focus groups.

## Testing

There is lots of other testing that is needed, besides just making sure that the product works.

**Useability testing** Useability testing is to ensure not just that the product works, but that it is ergonomic and easy to use. There are too many examples of products that could have done with some useability testing before they were released on to the market. For example, VCRs are notoriously difficult to program, and my cell-phone has facilities that I have never found, let alone used, and other features, like badly played tunes for ring tones, that I wish it did not have.

‘THERE ARE TOO MANY EXAMPLES OF PRODUCTS THAT COULD HAVE DONE WITH SOME USEABILITY TESTING BEFORE THEY WERE RELEASED ON TO THE MARKET’

Useability testing is typically done with either a specially instrumented version of the product, or by videotaping people using the device to do set tasks, along with questionnaires and the usual focus group mechanisms. The object is to see how the subjects use the product, and whether it can be made easier, either to use or to comprehend. What do they find difficult, and why? Maybe the instruction manual needs revising, or the help pages need expanding. Perhaps the user interface metaphor is not used consistently, or could be improved. Maybe the menu items could be moved around, or re-ordered, or additional short-cuts provided.

**Market testing** We discuss market testing in detail in the next chapter. You will not only need to market test the product, but also the advertising and other marketing collateral, as well as distribution and maintenance mechanisms. It's always a heart-stopping moment when the first helpdesk calls come in, as real users will do things a bit differently from the way you were expecting.

**Standards and regulatory approvals** You may need to have the product tested to get certificates of conformity to various standards, and regulatory approval for things like connection to the phone network, safety and emissions. These can take some time and expense to obtain, and may prevent the product from being modified without retesting. However, the product often cannot be legally sold without such certification, and therefore such tests are firmly on the critical path. Each country has its own set of standards, approved test laboratories, and scale of fees. Although there are some moves to mutual recognition and international standards, you should reckon on having to get the product certified for each territory. Some standards, such as the CE safety mark, are self-certified, but it is still wise to have the product independently tested in case of dispute.

Hopefully, you will have designed the product bearing in mind the standards to which it must conform, as it can be hard to retro-fit. For example, UK electrical safety standards require that any external screw must be able to be replaced with a screw five times the length of the original without coming into contact with a live circuit. This criteria can be hard to achieve, unless it has been designed in from the start. Some requirements, such as some telecom standards can be met by using components with existing approvals.

Medical and drug testing regimes are a world of their own. They are very long-winded, and conformance often dominates the cost of new development.

## Maintenance

We discuss maintenance below. A whole organization needs to be set up to deal with the maintenance issues of a product, which will require its own staff recruitment and/or training. Taking just one aspect of the maintenance operation, such as the helpdesk, which could be either in house or subcontracted, you will need to train staff, write answer scripts and FAQs (answers to frequently asked questions) and implement escalation procedures for when the front-line person can't resolve the query.

One thing that usually gets forgotten is what to do with the returns. With the best will in the world, no manufacturing and distribution system can be perfect, and besides, people are just plain awkward. You will get a percentage of your goods being returned, either because they are broken or have been broken, or because the customer has done something inconsiderate like dying or changing their mind, or maybe because your shipping department just got the wrong address. You can expect return rates of anything from 10 to 50 per cent if the product is at all complex or marginal. Sinclair Electronics, one of the pioneers of home computers, had 90 per cent returns for some devices. If you are shipping 100K per annum, 10 per cent is something like 1,000 per month, or 50 a day. It's when you see them stacked up in the post room, some bust, some not even opened, that the issues of mass market hit you. You need to set something up to return the good ones to stock, repair the repairable, deal with the customers who have sent theirs back, and find out if there is anything you can change that will reduce the number coming back. If you are manufacturing overseas, the problem is worse, as the manufacturer will typically operate a return to factory warranty, and that involves even more material movement and administration. It may be cheaper just to trash the broken ones and send the customer a new one from stock.

❛ ONE THING THAT USUALLY GETS FORGOTTEN IS WHAT TO DO WITH THE RETURNS ❜

For web pages and information products, the problems are related but different. Clearly, such services need maintenance, helpdesks and customer service organization too. You also need to find out how people use the services, and where the effort needs to go in improving them. It is likely there will be some parts, such as the home page, that get a lot of traffic, and others that see no traffic at all. A certain amount of insight can be gained from the click record, although analyzing it is quite difficult. Also, it won't tell you why, for example, a customer got part way through entering an order and then gave up. Was the page too complicated? Did the telephone ring? Was the product too expensive, or the wrong colour or size? The only real way to find out is to ask a sample of the users. However, organizations are strangely reluctant to do this, maybe because they will almost certainly not like the answers, even if they need to hear them.

## Exercise

**one** Take a product you are familiar with, such as a VCR or a mobile phone, and brainstorm how it could be improved to give a better quality of user experience.

## Further reading

Moore, G.A. (1998) *Inside the Tornado*. Capstone

Moore, G.A. (1999) *Crossing the Chasm*. 2nd edn. Capstone

Rogers, E.M. (1995) *Diffusion of Innovation*. 4th edn. Free Press

# 12 Quality and maintenance

‘The quality of mercy is not strained
It droppeth as the gentle rain from heaven
Upon the place beneath: it is twice blessed:
It blesseth him that gives and him that takes.’

(William Shakespeare, *The Merchant of Venice*, 1596–8)

THIS CHAPTER IS ABOUT THE SEEMINGLY BORING BITS that sometimes get forgotten in the rush to get something new out of the door. They are not boring at all, but are in fact the real foundation of the business.

They are:

- quality
- maintenance
- documentation.

We look at each in turn.

## Plan for quality

This section is about the need for quality, and some of the things that you can do to build it in. Quality is hard to define, but it's one of those things that you're sure about when you come across it. Read *Zen and the Art of Motorcycle Maintenance* (Pirsig) for a classic treatise on quality.

> QUALITY IS REMEMBERED LONG AFTER THE PRICE IS FORGOTTEN
>
> (GUCCI FAMILY SLOGAN)

Quality is what gets you the second sale, and stops the customer bringing the first sale back for a refund. Although Gresham's Law states that the mediocre drives out the good, and shoddy goods can always be made cheaper, quality is always better in the long run. It is the quality of your products that keeps the customers coming back, and lets you build a stable business for the long term.

> IT IS THE QUALITY OF YOUR PRODUCTS THAT KEEPS THE CUSTOMERS COMING BACK

## Gresham's Law

Sir Thomas Gresham (1519?–79) was an English merchant, diplomat and financier in the reign of Elizabeth I, a time when there was considerable clipped and debased coinage in circulation. His name was given to Gresham's Law, the economic principle that 'bad money drives out good', that is people prefer to hoard the good-quality coins, and pass on the bad, and so that the general coinage in circulation suffers.

Quality is not something that can be bolted on, but must be integral to the product and the development process from Day One, and must be deeply embedded in the corporate culture. The person responsible for quality assurance (QA) in the company needs to be at Board level, as considerable authority will be required to stop products from being shipped that don't measure up, in the face of an impending deadline.

## Three dimensions of quality

Quality means at least three separate things, all at the same time:

- feature set
- process
- reliability.

**Feature set** Feature set is an aspect of the quality of the design. Does the product do what it promises to do? The user has bought the product to solve a problem of theirs, or serve a particular purpose. Therefore it is paramount that the product can indeed meet these criteria. The product should do what it says it will do, and should do it without any nasty surprises, unexpected consequences or side effects. It should be easy and natural to understand, and also efficient, economical, ergonomic and even elegant.

Achieving these things is the aim of good design, yet defining the qualities of good design is much easier that trying to explain how to achieve it. However, you will know when you have come up with a good design, as it will just feel right, and will have a simplicity about it that makes it seem effortless. Good design also makes things easy to test and to manufacture.

Nowhere is good design more important than for web pages. Web pages are your shop window, the external face of your enterprise for the world, and it is therefore essential that they are well designed. Look at **www.webpagesthat-suck.com** for real-life examples of what can, and does, go wrong.

**Process** Having the right processes in place is important not only for the design phase, but also for manufacturing. Can you produce the product on time, every time? Much quality assurance is about reproducibility, and includes ensuring that the same level of quality and detail is achieved in all products, and that the design achieves its aim. The objective is to lessen the risk of things going wrong. We discuss this in more detail below.

> NOWHERE IS GOOD DESIGN MORE IMPORTANT THAN FOR WEB PAGES

**Reliability** Reliability can be defined as the rate at which a user will encounter anomalies. Unreliable products are expensive to support and maintain, besides giving you a bad reputation. Product recalls are a nightmare. Testing helps, but testing can never be complete, and there is no substitute for making sure that the product is fault-free in the first place. However, if a fault is discovered, it is important to ascertain how far its effects will spread, in order to gauge the appropriate scale of counter-measures.

## Risk

You cannot armour yourself against every eventuality or problem that might befall you, as you have only limited resources. However, you can take measures to ensure that your resources are deployed in the most efficient manner, and so give yourself the very best chance of success. So how do you decide where the effort should be put for maximum effect? The first thing to do, is to perform a risk analysis. For instance, if you are assessing your security policies, you will first need to consider the varying effects of the different security threats faced by your company. The mere existence of a theoretical threat does not mean that the counter-measures would be cost effective. For example, counter-measures against earthquake damage in southern England are probably not cost effective, even though there is a small possibility of occurrence. Employees may steal the occasional paper-clip with little ill effect, but complete loss of all computer data would be business-critical.

The risk is composed of:

- the *threat* (for example a hacker intercepting an e-mail) with a certain probability of occurrence
- the *vulnerability*, which is the probability of the threat being successful, and
- the *value* of the asset being protected.

‘COUNTER-MEASURES AGAINST EARTHQUAKE DAMAGE IN SOUTHERN ENGLAND ARE PROBABLY NOT COST EFFECTIVE’

We can now define the risk as the multiple of the three values:

$$\text{risk} = \text{threat} \times \text{vulnerability} \times \text{value}$$

People are notoriously bad at estimating risk and probability – just think of how many people participate in things with a low probability of success, like lotteries. Performing a formal risk analysis – drawing up a table of risks and ranking them in order – is a good discipline for many aspects of enterprise, and will help you to deploy your resources to where they are most needed.

## Standards

The dominant standards for quality assurance are the ISO 9000 series, with ISO 9000:2000 being the latest version. Key to an understanding of these standards is the notion of *traceability*.

The ISO 9000 standards originally arose from a military need for the guaranteed safety and interchangeability of equipment. This need was prompted by incidents such as the tragic case, in the 1960s, of a Harrier jump jet that exploded on take-off. Subsequent investigation showed that bolts holding together the jet's engine casing were faulty, and failed under load. The identification of where all the other bolts from that particular faulty batch were became a matter of urgency.

Implementation of ISO 9000 requires knowledge of where everything came from, and where it is going. You cannot just chuck a new supply of, say, bolts into the drawer with the last of the previous batch, as you need to keep each batch separate, with documentation of their origin, right back to the raw steel ore. The calibration of all test equipment needs to be referenced to some national standard. It's a lot of work, but vital in certain industries.

For design and software, each decision needs to be traceable, with supporting documentation giving the reason for that decision, and review processes to check the decision. There needs to be a system of control, defining who can alter what document or piece of code, and audit trails recording who altered what, when, and why. Nothing in the design, code base or any other controlled document gets altered without the alteration being reviewed, authorized and checked.

> 'FOR DESIGN AND SOFTWARE, EACH DECISION NEEDS TO BE TRACEABLE'

Other standards that you should be aware of include the ISO 14000 series of environmental management standards, and BS 7799/ISO 17799, which is the code of practice for information security management. In addition to these national and international standards, companies and their purchasing organizations may also have their own internal standards for things like documentation, code style, and so on.

## Control and review

Central to the notion of quality assurance is the need to record key decisions, and to review them regularly.

For traditional waterfall decomposition, a review is required at the completion of every level of decomposition, from the design documents to the specifications of the interface of the modules. Such reviews are done by design walkthroughs, or by emulation.

**Key decisions and key documents** A software project will usually have, as a minimum, the following controlled documents:

- project definition
- user requirement document (URD)
- project constraints document (PCD).

These documents, or their equivalent, are an outline definition of the project from the customer's point of view. The URD says what is to be built, and the PCD says how it is to be built. The URD is typically produced by, or in conjunction with, the customer. It may include mock-ups and illustrations of, for example, screens, and sample input and output.

The URD specifies priorities. One way in which this is achieved is to use a system of priority markers, with, for example, statements marked 'A' as mandatory, statements marked 'B' as desirable and statements marked 'C' as good to have. More usually, however, semi-formal language is used, with statements including 'must' or 'will' defining the mandatory features, as in 'The system must respond within one second'; statements including 'should' or 'can' indicating desirable but not mandatory features, as in 'The system should respond faster than 0.1 second in normal circumstances'; and statements including 'may' or 'might' indicating additional features, as in 'The system may also include a holographic projector'.

The PCD covers things like budget, timescales, resources and any other constraints required, such as compatibility or the tool-set to be used.

**Base definition** The base definition is normally captured in the following:

- functional specification or prototype (FS)
- top-level design.

These two documents are produced by the project architect, and define the project further. They are written from a technical viewpoint.

The FS provides detail on *what* is to be built, explaining every screen, input, output and file structure, etc. The FS may also include the database design, if one is required. Some people include the first draft of the user manual in this specification, which is good practice, as it is wise to set controls for the manual from the start.

Long-winded functional specifications are often difficult to read and comprehend, and it is a good idea to accompany the FS, or even replace it, with a prototype, at least of the user interface. This will help the client to have a better understanding of the project. The FS needs to be verified against the original URD, both by design walkthrough, and by checking each statement in the URD against its implementation in the FS. In turn, every statement in the FS generates a compliance test against which the eventual product is tested.

‘LONG-WINDED FUNCTIONAL SPECIFICATIONS ARE OFTEN DIFFICULT TO READ AND COMPREHEND’

The top-level design document says *how* the product will be built. It includes a description of the overall architecture, and detailed specifications of the top-level modules, key data structures and the interfaces between them. It is from this document, in the traditional model, that the process of waterfall decomposition starts. In the non-traditional model, the top-level document sets out the process of incremental development, describing in what order things should be built. The top-level design document is verified against the FS by design walkthrough. Thorough checking at this stage is very important, as it is hard to turn back once you have started down the wrong path.

**Change control** Change control needs to be implemented to ensure that the consequences of a change are reviewed properly before the change is made. In addition to the documents above, the following documents will also normally be part of change control:

- project plan
- project log
- quality plan

- documentation plan
- sub-system specifications and interfaces
- data model and dictionary
- module specifications and interface
- released code and documentation
- deliverables
- test results.

## Versioning

A useful strategy to use in change control is versioning. Most automated source control tools allow different versions to be produced. Versions allow you to roll back to the last known successful version, if all goes wrong. Versions might include:

- current test verion
- last released version
- versions for specific applications
- versions for different environments.

**Definition of standards** Internal standards that need to be defined include:

- coding standards
- naming conventions
- routine structure
- testing procedures

- documentation standards
- house style
- conventions and examples.

**Process review and audit** It is no good putting these standard processes into place, and just sitting back and hoping that everyone will automatically use them. You can't expect things to run smoothly by themselves. The processes will need active management, and regular review. More importantly, you should audit your processes at regular, but random intervals, to ensure that they are being complied with, and that, for example, all documents have been properly reviewed and signed off.

## Plan for maintenance

### The relationship continues

Your relationship with the customer does not end when you ship the product. You need to keep in touch with your customers, listen to them and ask them questions to enable you to target your products more closely to their needs. Maintaining a good relationship will involve providing services such as aftercare and helpdesks.

The revenue that can be gained from maintenance services can be worth far more than the original capital cost of the service. Maintenance service revenue is ongoing, and typically equates to about 15 per cent per annum of the original capital cost of the contract.

“THE REVENUE THAT CAN BE GAINED FROM MAINTENANCE SERVICES CAN BE WORTH FAR MORE THAN THE ORIGINAL CAPITAL COST OF THE SERVICE”

Of course, maintaining a good relationship with your customers has other advantages as well, such as providing a proven channel for future sales, and an inside track for future procurements. For this reason, your maintenance engineers are often your best sales people.

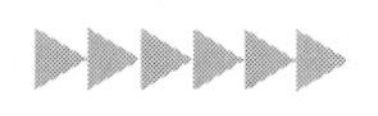

Frequencies of fault are often measured as mean time between failure (MTBF). As a product matures, this value should drop as the bugs are fixed, but will climb again at the end of a product's life.

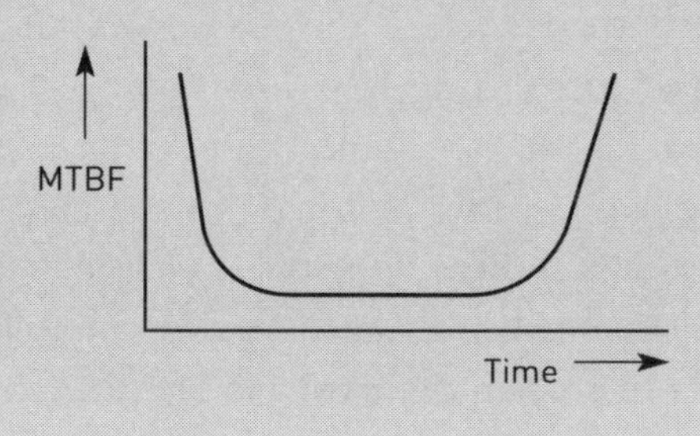


## Levels

The maintenance required is dependent on the task, and the level of service needed. It can range from telephone support during office hours, to 24/7 on-site cover for critical applications. Parameters include:

- Hours of operation.
- Minimum response time. Note that this is response time, not time to fix.
- Mean time to repair (MTTR). Often not specified.

## Helpdesk

The helpdesk is the first line of defence for most maintenance calls. Helpdesk staff deserve medals, as working in a call centre is the nearest thing to slavery that the law allows. They are also first in the line of fire, and have to deal with irate customers when things go wrong. Helpdesk urban myths abound (search the web for 'helpdesk stories'), such as the helpdesk engineer who told a user to close all open windows. The user did. Throughout his apartment.

Setting up a helpdesk is hard work if you are expecting any volume of calls, and you might be wise to outsource the helpdesk provision to a specialist sub-contractor.

Helpdesks usually work to scripts or FAQs (answers to frequently asked questions) that are often implemented as a database of answers. If the helpdesk agent can't find the answer to a particular question, they trigger an escalation process – first consulting their supervisor, who is (hopefully) more experienced, and then an off-line engineer.

Writing the helpdesk scripts and training the initial agents, who can then train further agents (cascade training), is an important part of launching a product. It is important to set up a reporting system to enable you to monitor the operation of the helpdesk. If you have outsourced the work, your contractor should have a reporting system already set up, otherwise you will need to set up your own, which might include the following statistics. The report does not need to be longer than one page:

- How many calls has the helpdesk answered?
- What is the average wait time?
- What is the number of calls that are dropped?
- What is the number of calls that are resolved satisfactorily?
- What are the top ten complaints/suggestions?

Calls to the helpdesk are often the first notice that you will get of something being wrong, particularly for an online service. The irony is, that if there is a major fault, there is no way that the helpdesk can keep up, and in such situations, many helpdesks just stop answering complaints altogether until the fault is fixed, rather than have their staff abused. However, when things do go wrong, the one thing that users need and want is information. An accurate recorded message, changed frequently, say every 15 minutes or so, goes a long way to help.

“WHEN THINGS DO GO WRONG, THE ONE THING THAT USERS NEED AND WANT IS INFORMATION”

Intelligent phone systems with voice or touch-tone recognition and direction can help cut down the number of calls .

There may be some corporate cultural issues which you will need to confront in order to effectively run a helpdesk. Sometimes, high-tech companies regard calls to the helpdesk as a nuisance and waste of time, and consider that the helpdesk is there to keep the public away from the 'real' people. Even worse, the people who make the calls are often considered as ignorant and stupid. ('If it was hard to build it should be hard to use!') Such attitudes are arrogant and will not help you or the customer. Calls to the helpdesk provide invaluable feedback about how your product is actually performing in the field, and how it can be improved. The helpdesk should not be a distant part of the organization simply doing its best to protect the company, but intimately bound in with the provision of service. Ideally, every member of the development team and their management should serve their turn on the helpdesk, at regular intervals.

❛THE HELPDESK PROVIDES CONTACT WITH REAL CUSTOMERS AND REAL PROBLEMS❜

The helpdesk provides contact with real customers and real problems. When a customer contacts the helpdesk, you have an opportunity to turn an unhappy punter into a real friend, and even to cross-sell them an additional or new product.

## Internal documentation

Key to maintenance is good internal documentation. Remember that you may have to maintain the product long after the original developers, and the knowledge that is in their heads, have left the company. There are many sad tales about companies having to recover listings from the engineer's children, who were using the last existing copy as scrap paper, or of having to ring up former employees in the hope that they will still remember something about the workings of a product that has just been returned for updating after many years. You have been warned.

## Plan for documentation

The job is not done until the paperwork is completed. Documentation is vital, otherwise your users cannot understand what they have, your staff cannot maintain the system, and your developers will keep going round in circles.

Good documentation takes a lot of effort – maybe three times the original coding effort – and it doesn't just happen by accident. The quality of documentation can result in part from the company's culture. For example, some companies develop house coding styles that encourage at least as much explanatory comment as code, together with standard format headers that explain what each sub-routine or module does, and how to call or invoke it. If designed well, some of this information can be mechanically checked, for example by processes that refuse to update the code base unless the code has at least as many comment lines as code, and is accompanied by the appropriate internal documentation. However, only human checking can ensure that the comments are actually useful and make sense, unlike the infamous comment in the sixth edition of Unix (circa 1975):

> *You are not expected to understand this.

(David Ritchie at **http://cm.bell-labs.com/cm/cs/who/dmr/odd.html** gives a full explanation of the origin of this apparently arrogant statement.)

Internal technical documentation and code comments are a good task for the co-pilot in a chief programmer team. Good documentation is one of the things that divides professionals from amateurs. Professionals are used to working in teams, where communication is paramount, while amateurs, and even academics, often work alone, and tend to use lower documentation standards.

> “GOOD DOCUMENTATION IS ONE OF THE THINGS THAT DIVIDES PROFESSIONALS FROM AMATEURS”

Writing good, clear documentation is a specialist skill, and it is unreasonable to assume that people who will be good technically will be good at writing too. To write good user-level documentation, you also need to approach the problem from a user's point of view, which is something that people who already have a deep knowledge of the design can rarely achieve. User-level documentation should be written with input from the team, but by a separate specialist, whose technical ignorance is one of their more valuable assets.

## Classes of documents

Documentation comes in a number of levels, specifically designed for different audiences.

**User documentation** This is documentation for the customers or users of the product or service. A typical set of such documentation would include:

- quick start guide
- unpacking and installation instructions
- user guide – narrative
- user guide – reference document
- FAQs
- help topics and system
- examples, guided tours and demonstrations
- video demo
- support web pages
- marketing collateral – adverts, press releases, leaflets, flyers, in-store demos, tent cards
- exhibition material and give-aways.

**Training material** The next set of documentation is training material, aimed for people like helpdesk staff. There is a particular type of training documentation that is specifically designed to teach people how to train others to use the product or system. This is known as cascade training documentation.

There are likely to be a number of different types of courses within this category. These should cater for different levels of understanding, for external and internal people, and for people with different aims, such as sales staff or maintenance engineers.

Such material might include:

- course material
- self-study material
- worked examples
- tests
- video presentations, etc.

**System-level documentation** System-level documentation is for the techies who will have to install and maintain the product or system in the field. It is technical documentation, but it may also be released to selected customers. It should be formally structured, with a comprehensive overview and reference sections. Such documentation may have associated training courses and web-based backup pages such as bulletin boards, newsgroups, mail lists and patches to download and support pages. Encourage your field engineers and system-level customers to share information, and build a community. If one of them finds a fix to a problem it may help others.

**Maintenance documentation** Maintenance documentation is the internal documentation that explains the situation as it really is. It includes all the original project documentation, the code base and its associated commentary, as well as specially written overviews and guides. Keep it safe.

## Avoid forward references!

One of the things that makes technical documentation particularly hard to read is the use of forward references, that is, explanations using terms and jargon that have not yet been explained. This can be hard to avoid, as the documentation needs to be written by people with an in-depth knowledge of the system, and such people will use, inevitably, the technical language that they are familiar with. Knowledge is not linear, and trying to force it into a linear format, such as a book or a help manual, can result in such effects. Non-linear media, such as help scripts or web pages with links, can be more effective.

## Conformance

The biggest problem with documentation is ensuring conformance, that is, making sure that the system actually behaves the way the documentation says it should, and that the documentation is consistent.

“THE ONLY WAY TO ACHIEVE CONFORMANCE IS TO FOSTER A DEVELOPMENT ENVIRONMENT AND CULTURE THAT UPDATES THE DOCUMENTATION WHENEVER A CHANGE IS MADE TO THE CODE BASE, AND TO TEST FOR CONFORMANCE AT REGULAR INTERVALS AND BEFORE EACH RELEASE”

The only way to achieve conformance is to foster a development environment and culture that updates the documentation whenever a change is made to the code base, and to test for conformance at regular intervals and before each release. It is amazing how easy it is for discrepancies to creep in. If a discrepancy is found, you will then need to decide whether the code or the documentation should change. This will depend, to some extent, on the relative effort involved. If the documentation is in machine-readable form, then changing it is not too bad. If it is printed, errata sheets or web page updates can cover minor amendments, but a major change is a big and expensive exercise. This can be a good reason to have the documentation in soft format, or at least in ring binders.

Finally, remember that a feature is simply a documented bug.

## Communication skills

Whether you are preparing documentation, presentations or web pages, good communication skills are vital. Here are some pointers for how to communicate your ideas clearly and well. Although they may be obvious, they are all too often ignored.

**What does the target audience know?** You can only explain things to people in terms of what they already know, and what they are interested in. It's no use going into deep technical detail for an audience of investors and bankers – their eyes will just glaze over, and they will drift away. Keep your explanations relevant to what they can comprehend, otherwise it will be like trying to explain colour to a blind person.

Your ability to communicate your ideas clearly and relevantly will affect how quickly you can get new ideas across. Don't talk down to your audience either. Although they may be ignorant about the things you are trying to tell them, which is why they are listening to you, they are not stupid. As the TV director said, never underestimate your audience's ignorance, but never underestimate their intelligence either.

If you are explaining big, new, complicated ideas, like your latest project, try not to cover more than one, or at most two, concepts per session, chapter or whatever. Beyond that your audience will hit saturation point.

And remember KISS – Keep It Simple Stupid. You should be able to explain your ideas to anyone. How would you explain what you are doing to an elderly relative? Have a go at trying to express your idea in no more than 25 words. This will force you to focus on the essential points.

On the other hand, it is possible to be too brief. Techies often tend to write in a very compacted style, cramming in facts and ideas, but arty people can be too verbose, producing yards of content-free waffle. The happy medium is somewhere in between. You also need to consider the structure of a presentation. Just as school essays should have a beginning, middle and end, you should aim to say everything three times; once in a summary in the introduction, once in the body of the text, and a third time in the conclusion. So, summarize what you are going to say, then say it, and then conclude with a synopsis of what you have said.

**Simplicity** Simplicity is also the key to presentations. If you are using slides, make sure that you don't put more than three to five points on each at most. The most popular software at the moment for creating presentations is Microsoft Powerpoint. However, don't get carried away by the technology, and fill the presentation with fancy dissolves, clip art and other giszmos. Marshall Macluan may have been right when he said that the medium is the message, but you want your message to be about what is projected, not how fancy the tools are.

It doesn't all have to be serious. Falling off the podium and telling a joke about halfway through the presentation will wake your audience up, and lighten the tone. However, be sure not to do this too regularly, as it can be counter-productive.

**Human factors for websites** The same principles of good communication also need to be applied to websites. We have already mentioned the excellent **www.webpagesthatsuck.com**, which attempts to teach good web design through demonstrating real-life examples of bad sites. The WebTV site **http://developer.webtv.net/design/** also has many wise words on how to set up a well-designed website.

‘HOW WOULD YOU EXPLAIN WHAT YOU ARE DOING TO AN ELDERLY RELATIVE?’

As for other forms of presentation, the main rules for good communication on websites are:

- Don't crowd it– include no more than three points on each page.
- Make it obvious and clear. Remember KISS – Keep It Simple Stupid.
- Use a deep tree of pages rather than a wide one. Do not use more than five hyperlinks per page, except in special circumstances, such as pages of links. Even then, five links per page, with a good commentary on why that particular link is useful, would be better.
- Avoid too many technical gizmos, animated icons and separators, and too much background music, etc.
- Aim to fit on the screen. Remember that not everyone has fancy monitors. Use relative rather than absolute positions and co-ordinates where possible.

- Use eye-friendly colour combinations – blue text on red background is never a good idea, neither is light grey on dark grey.
- Make each page useful. Front pages that just show a single icon or only fancy animation deserve the derision they get.

## Exercise

**one** As it says on the publisher's website:

'Try taking the "25 words, five messages and a question" test:

Our sales teams often have to pitch new books to booksellers in as little as 30 seconds. Copy space in retail catalogues and in-store systems is very limited indeed, and if you can summarize your pitch in 25 words or less, we will be able to sell it more effectively. Next, list the five most compelling messages that your book will contain; bullet-points will do. Finally, imagine that you are about to make a presentation to a large auditorium full of potential readers of your book, be they managers, students or investors. What one question would you ask them to engage their minds, and make them realize that this was something both relevant to, and compelling for both them and their businesses?'

## Further reading

Pirsig, R.M. (1974) *Zen and the Art of Motorcycle Maintenance*. Vintage

ISO 9000:2000 is available from **http://bsonline.techindex.co.uk**, as is BS 7799-2:1999, *Information security management. Specification for information security management systems*

# Marketing and selling

# 13 Reading the market

> Marketing is the key factor in business. It is not only the fuel, it is the compass of the ship.
>
> (Sir John Harvey-Jones, former chairman of *ICI*)

THIS CHAPTER AND THE NEXT ONE ARE ABOUT MARKETING. Marketing is quite different from selling, even if they are normally lumped together. Sales and marketing require different skills, different mindsets and different types of people. It is rare to find someone who is good at both.

Sales and marketing are not second-rate activities, but are the primary engine of the company. They generate the cash to fuel the enterprise and keep it running, and they are the people who deal with the customers daily and directly. They are the public face of the enterprise.

## Sales and marketing are different

'Marketing is the management process responsible for identifying, anticipating and satisfying customer requirements profitably', according to The Chartered Institute of Marketing (**www.cim.co.uk** ).

Marketing is deciding what to sell, for how much, to whom and by what channel. Marketing is also about customer communications, building an image for the company, and spreading the news about your products through methods such as advertising, mail-shots, PR, and websites. Marketing is mostly a cerebral activity, done from your desk.

Selling is about actually moving the product, and dealing directly with the customer. It is much more of a contact sport than marketing, and needs people good at building relationships, rather than the more analytic marketing approach.

We look at marketing first, and sales in Chapter 15.

## What should you be selling?

One of the main responsibilities of marketing is the decision of what to sell, and how much to sell it for. We looked at market research, which is one of the major tools, in Chapter 2.

## Product or service requirements

Unfortunately the old adage about 'invent a better mousetrap, and the world will beat a path to your door' is not true in the slightest. For a start, mice, or at least the sort of mice your mousetrap is designed for, might be extinct or no longer a problem.

For people to buy your product, they must:

- *Know about it*. The customer must be aware of the product in order to purchase it, either through advertising or word of mouth or some other means.
- *Have the opportunity to purchase*. Not only must the customer know about the product, but they must also be able to buy it. This used to involve the manufacturer negotiating for shelf space in the retail distribution chain, but now 'having the opportunity to purchase' can be achieved by the manufacturer running a website from which the goods or services can be ordered. The retail route is still important, and even for websites there are the same problems of how to attract traffic to the sales point. The web equivalent of competition for shelf space is space for links on popular high-traffic sites, and premier positions in search engine listings.
- *Consider it to satisfy a real or perceived need*. The user needs to be convinced that they will benefit from making the purchase. The benefit need not be tangible, but can be some intangible value.
- *Consider it affordable (but not too cheap)*. Part of having the opportunity to purchase is being able to afford it. On the other hand, if it is too cheap it may not be valued, or suspicion may be raised that there is something wrong.

‘THE USER NEEDS TO BE CONVINCED THAT THEY WILL BENEFIT FROM MAKING THE PURCHASE’

See also the ACCTO criteria in Chapter 11.

## Product characteristics

A function of marketing is to try and determine the optimum characteristics for the product, which can be quite hard. For example, suppose you were a PC manufacturer. How would you decide how much memory or what size of disk to put into your PCs? What other features, such as a DVD player, would you include?

Of course, it will be important to consider the opinions of your technical team when considering the range of features, as they will be able to advise on the technical implications and cost. However, more important will be the information that you will gain from market analysis which can provide data on trends in the marketplace and the positioning of competitors, feedback from potential customers and similar soft measures.

Company and brand policy also need consideration. Will the product be a low-cost, high-volume, cheap and cheerful product, or a high-cost luxury product with all the bells and whistles?

What makes your product or service stand out from its competitors? You need to identify these 'unique sales points' (USPs) and ensure that they are attuned to what the market wants. In Chapter 1 we introduced the features, advantages and benefits (FAB) analysis, and remarked that what sells are the perceived benefits, even if these are intangible. You need to analyze each of your USPs for the benefits it gives the customer. If the primary benefit is intangible, or brand value, then you need to consider what you need to do to support and maintain the intangibles. Perhaps you have to educate the customer as to just what the benefits are. Maybe it's high spending on advertising, or scarcity value, or like Rolls-Royce used to be, just phenomenal build quality. In the service sector, like Schlumberger, you might want a reputation for always getting the job done, no matter what, or to build relationships with your principle customers so that you become an extension of their company, and hence the natural first port of call for any new work.

## Price sensitivity

There are many factors that contribute to the decision of what price a product or service should be sold at. Price is relative to the perceived value, but is the value due to its utility, its intrinsic value, or its apparent value? Intrinsic value relates to its actual cost. Utility value relates to its usefulness, and perceived value relates to what the user thinks it's worth.

Consider a piece of gold jewellery. The gold that it is made out of has a certain intrinsic value. Fashioning the gold into a piece of jewellery will cost a certain amount in labour hours and facilities, but it will probably increase the value of the object by more than the cost of the process, depending on the skill of the craftsman. However, if the piece of jewellery is made by, or associated with, the name of a well-known artist or a designer brand, something magic happens and the price increases dramatically, due to the perceived value of the brand or name. Here the price is governed by many intangible values, rather than the intrinsic cost of the gold and the craftsman's wages. These intangible values might include the self-esteem of the buyer, which is boosted by owning a piece of designer ware, and the implied lifestyle aspirations that go with it.

Digital watches provide another interesting example. Two nearly identical watches can vary dramatically in price, even though they utilize exactly the same electronic movement and materials. There is a market for extremely expensive mechanical watches, which are marketed as high-end luxury goods. Such watches can cost up to 1,000 times more than a standard model (some were recently advertised for prices in excess of $5 million), but actually do not perform as well at telling the time. The reason that they sell is because of the perceived value of the brand and the status associated with it.

‘DIGITAL WATCHES PROVIDE ANOTHER INTERESTING EXAMPLE’

## Pricing strategies

Pricing strategies can have a number of different bases, but are based on two main dichotomous strategies: supply-side strategies which are broadly about what it costs to make the thing plus a margin for profit, and demand-side strategies which are about what people will pay for it. Although prices are set by a supplier, they are governed by what the buyers will pay.

There are three main pricing methods:

- Cost based (supply side)
  - Breakeven analysis. How much does it cost to produce the goods?
  - Cost plus. Calculate cost then add some profit.
  - Marginal cost pricing. The cost of producing one more.
- Competition based. How much is the competition charging?
- Perceived value pricing. What does the customer think it's worth?

Supply-side pricing can be used to set a floor, that is a lower limit below which it is not worth selling the product, except in special circumstances. One such special circumstance could be if you are using the product as a loss leader, which is a technique often used by supermarkets and involves setting the price of various basic commodities below cost, in order to attract people into the store, where they might also buy higher-priced and more profitable goods while they are there. Other examples of loss leaders include mobile phones, where the cost of handsets are subsidized from the profits made on subsequent service agreements, and the current land-grab situation on the internet, in which the objective is to establish a dominant position in a niche market, driving out competitors in the hope that future revenues can be extracted from the near monopoly position.

Extraordinarily the land-grab mentality was actually encouraged, until recently, by the stock markets, due to the fact that internet companies have little or no profitability on which to base a conventional valuation, so valuations were often based on the number of customers. If lowering prices increased the number of customers, perversely the valuation of the company went up, despite the increased costs and reduced profitability. Fortunately some sense is now returning to this sector, following the bursting of the internet bubble.

## The economics of supply and demand

In a perfect market there would be a simple linear relationship between price and the number of products expected to be sold over a particular period. This is called the demand curve D(p). Prices can even be negative if people are paid to take the product. For the moment, we will assume that price and quantity sold in a period are linearly related. They probably won't be, but the principles involved are the same and it is easier to see what is going on if we make that assumption.

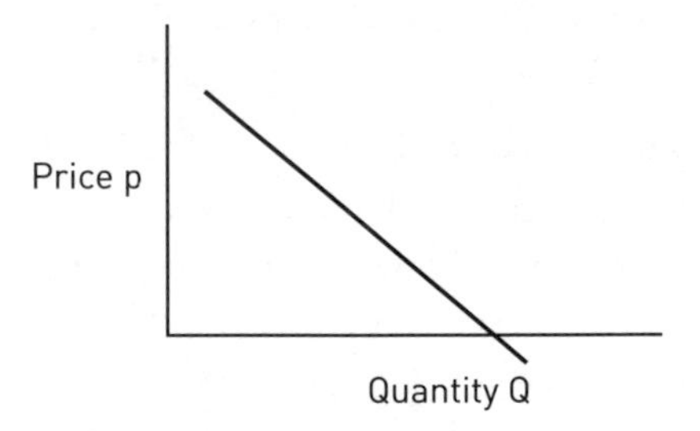


Goods must be available, and be able to be produced, at the time when the customer wants to buy them. Demand is said to be *elastic* when the demand responds to a change in price, and *inelastic* when a price change does not significantly affect demand.

Economists define price elasticity (e), as follows:

$$e = \frac{P}{\Delta P} \frac{\Delta Q}{Q}$$

where P and Q are a particular price and volume, e is the *elasticity* of the market, $\Delta Q$ the change in quantity sold in a period, caused by a change of price $\Delta P$. The demand curve is not always a straight line; that is, e is not always constant, and the price elasticity can change with price level, especially at very high or very low prices.

**Price discovery – supply** Related to the demand curve is an equivalent supply curve S(p). If the price in the market rises, more suppliers come into the market and more product is available. It's probably not a linear relationship either, but we will use a straight line as an approximation.

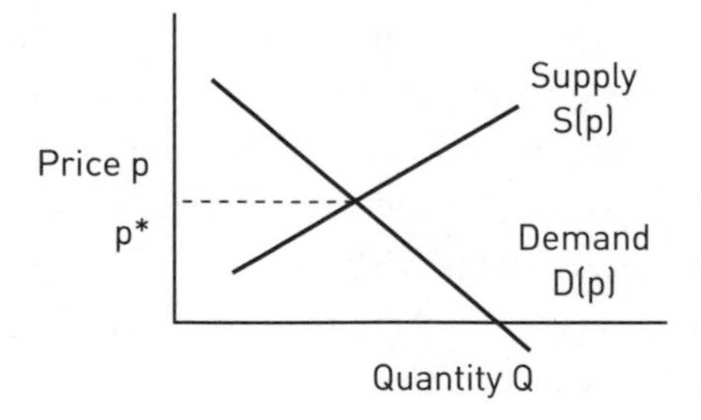


Classical economics teaches that, in the medium term, there will be a stable price p* and volume where the two lines cross. If the price is higher, more product will come on to the market than is consumed, the market is said to be *oversupplied* and manufacturers will cut prices to shift the product, moving towards the equilibrium. If the price is lower than the equilibrium point, there is more demand than product, the market is said to be *undersupplied*, and the price goes up. More realistically, because markets are not perfect, and there is delay in the feedback loop, the price and supply oscillate about the stable position, between glut and famine. If the supply and demand curves are non-linear, this oscillation can be chaotic in nature.

In the longer term, technology may change and may shift the supply curve to, say, S'(p), or governmental taxes or subsidies may shift the demand, and there will be a new equilibrium point p'*.

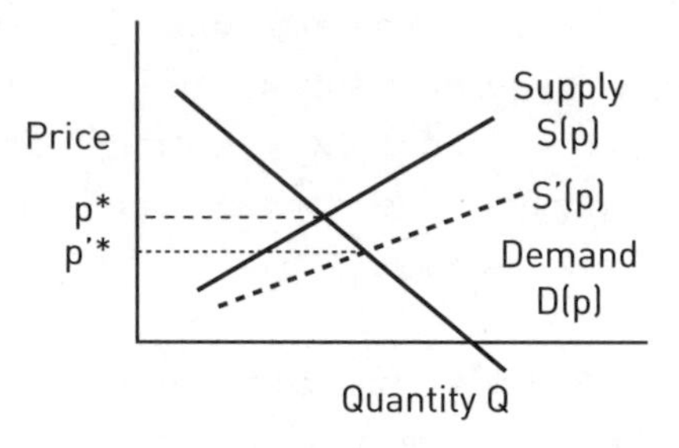


**Effect of economies of scale** In Chapter 5 we saw that cost of manufacture consists of fixed costs which have to be paid regardless of how many units are manufactured, and variable costs which are dependent on the number manufactured. The unit cost goes down as the quantity of products manufactured goes up, since the fixed costs can be amortized over a larger number of units. In addition, larger economies of scale allow the manufacturing process to become more efficient, and the purchasing process to drive harder bargains with suppliers. To illustrate the effect of economies of scale on cost, we can plot a manufacturing cost curve. Note that this is a different curve to the supply curve above – it is the cost of manufacture, rather than the selling price. To make things simple, let's assume a fixed cost of zero and a linear manufacturing cost curve, and also that things get cheaper as the volume goes up. This is not always so, especially if plant or machinery are near capacity.

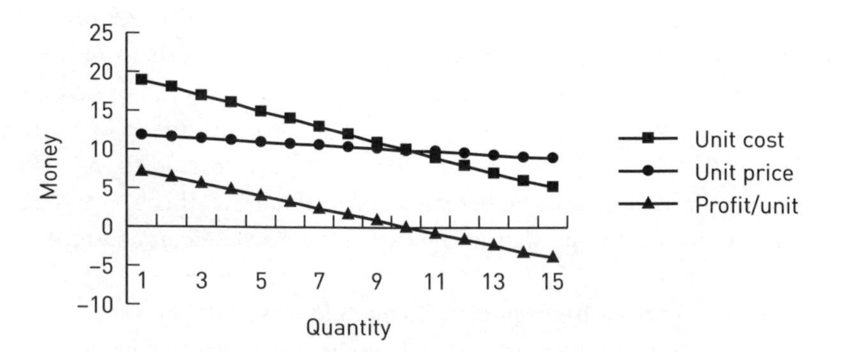

The distance between the lines is the profit per unit sold. The place where they cross is the *break-even point*, which is where the price is the same as the costs. If you are a buyer, this is the price at which you want to buy.

The calculation of the optimum selling price is a little more complicated. Since the gross margin is the selling price minus the cost, then multiplied by the quantity, then: GM= Q(P-C), where P is the sales price and C the cost of manufacture at that volume. However both P and C depend on Q. Therefore, the gross margin, GM, depends on $Q^2$ (a quadratic), and hence there is some price at which there is a maximum profit.

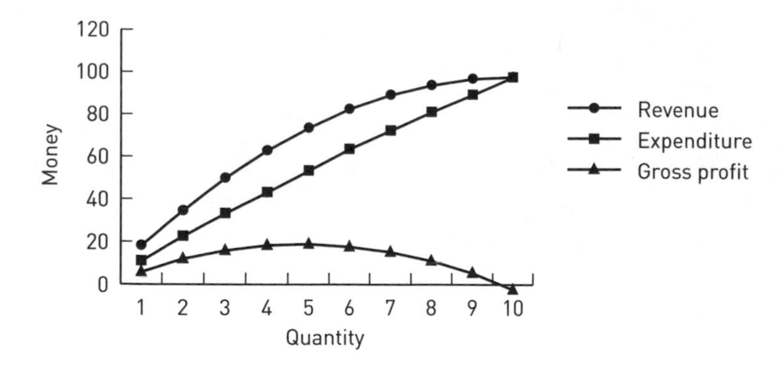


If the fixed costs are not zero, the supply curve is not a straight line, and a typical curve might look more like this:

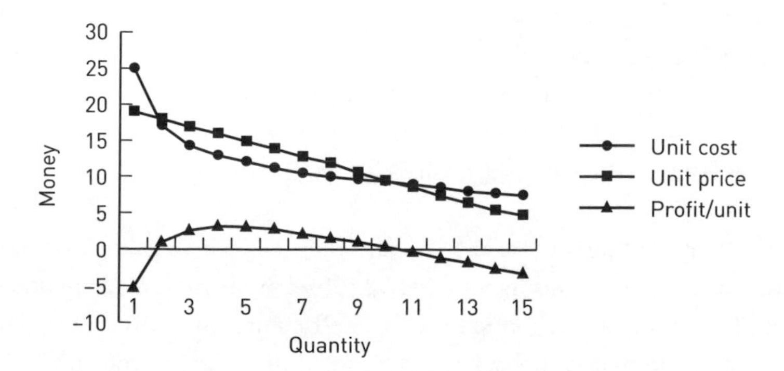

If we multiply unit price and cost by the quantity sold, we will get the total revenue and expenditure, and hence the total profit. Note that this does not necessarily occur at the same price as the optimum profit per unit.

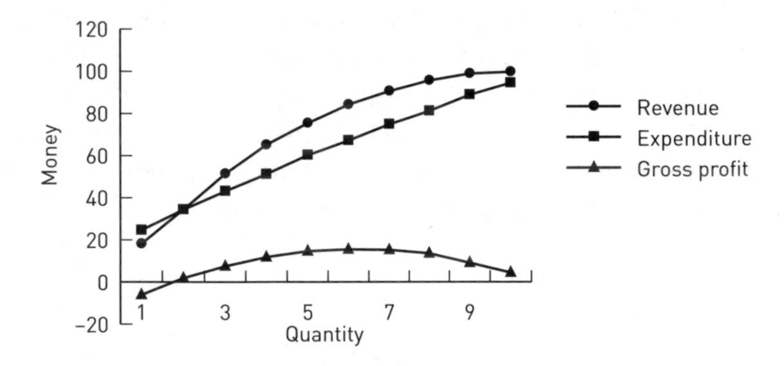


Look at the shape of the gross profit curve. Note that it is fairly flat, as the gross margin is relatively insensitive to small variations in price, and the decrease in revenue from a lower price is made up by increased value from increased volumes. However, this is not always so, particularly with small margins. Where cost and price curves are non-linear, there may be more than one maxima, for example a high-price low-volume and a low-price high-volume unit may be equally profitable.

On the whole, people price too cheaply. Most markets are relatively insensitive to small price changes, and the effect of a ten per cent price increase, if all other things are equal and the market is inelastic, will have a large impact on your profit, with probably only a minor decrease in sales. Since profitability is typically only ten per cent of the selling price, a price increase of ten per cent may immediately double the profits. Lower volumes are easier to manufacture as well, so there may be additional benefits in selling slightly less.

‘ON THE WHOLE, PEOPLE PRICE TOO CHEAPLY’

**Efficiency** A market is said to be Pareto-efficient if the allocation is one where you cannot make someone better-off without making someone else worse off. A competitive market is Pareto-efficient. An ordinary monopoly is not – the monopolist chooses a price to maximize the income, p*D(p).

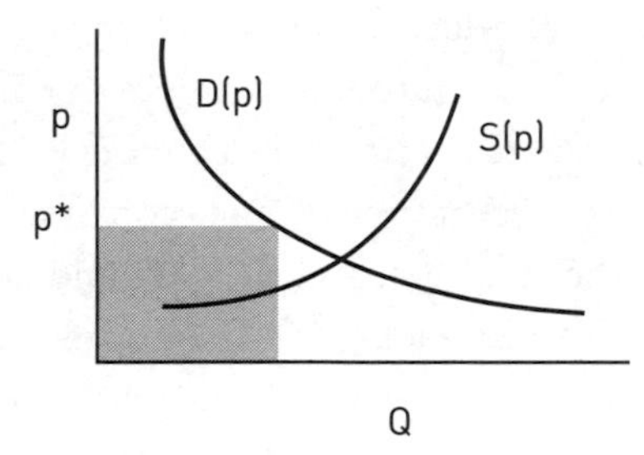


However a 'discriminating monopolist' will sell to everyone at just the price they are willing to pay. This is Pareto-efficient, as the same people get the same goods as in a competitive market, but additional people also get goods at prices they are willing to pay.

Examples of such pricing is where superficially different versions of the same product are created targeted at different market groups, for example airline ticket prices, or car models with small differences This is based on the psychology of the distribution curve, as when customers are given a choice of low-, middle- and high-end products they tend to choose the middle one. If you introduce a high-end product to your range the sales of the middle range product tend to increase. This is known as the 'Goldilocks effect'.

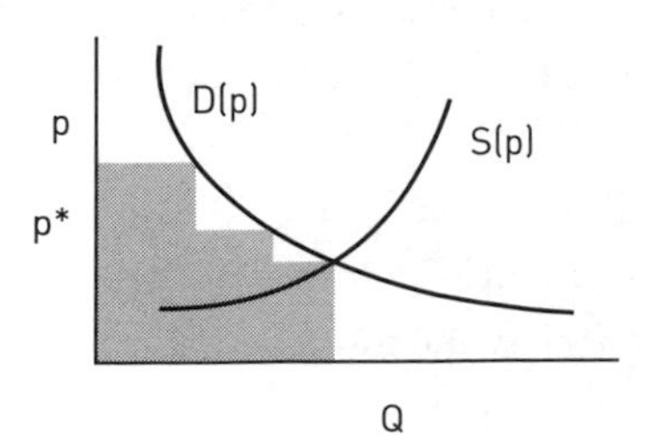


**Lock-in and network economics** For some goods and services, such as most software, the marginal cost of manufacture is close to zero, yet they sell for substantial amounts. Why doesn't market competition force down the price?

The answer is that some products create a lock-in effect. When you buy a product you may be committed to buying more or complementary products. If you buy a PC you are committed to buying PC software. Changing to, say, a Mac or a Unix-based system then becomes more difficult because of the investment in the software, and the data and documents that only that software will read. You are locked in. There are many other examples, such as the difficulty of changing your e-mail address or phone number, and examples where the lock-in is artificially created, such as frequent-flyer clubs and loyalty cards. In some cases the switching costs are borne partly by the supplier. Where the cost of goods or services are low, the economics will drive the net present value of the customer to that of the total switching costs, that is the total of the switching costs of the customer and the new supplier. Suppose you are an internet service provider (ISP), and the NPV of a new customer is, say, £100, in terms of discounted revenue. Suppose it costs the customer £50 in hassle to switch suppliers, and it costs you £25 to set up the new customer. The total switching costs are thus £75. You can offer the customer a £60 cash-back to join your service. They are thus £10 ahead, but you are £ 100-£25 -£60 = £15 ahead!

‘IT GETS MORE COMPLEX WHERE THE SWITCHING COSTS ARE ASYMMETRIC’

It gets more complex where the switching costs are asymmetric. For example a mobile phone company may need to give the switching customer a new phone, at a comparatively high cost, but the incumbent company can bribe the customer with a loyalty bonus of free minutes, at a low marginal cost.

An example of widespread lock-in occurs with network effects. Many networks have the property that the more people use them, the more valuable they are to each user. An early instance of this was the development of the telephone network, where until there were a critical mass of users the utility of a telephone connection was largely status. Other examples include the growth of fax machines around 1985, and the growth of e-mail around 1995. Networks do not have to be physical. They can be virtual as well, for example the number of people using a particular marketplace, like eBay, or using a PC. There is a snowball effect for the dominant player – people write software for a PC because there are so many of them, and people buy PCs because there is so much software available.

Metcalfe's Law states that the value of a network is proportional to the square of the number of users. This is not quite true – a network tends to switch from low to high utility as each group of users achieves critical mass. However it tends to mean that once a network passes a certain critical size, the winner takes all. These network effects are described by economists as 'external' factors.

A one-sentence summary of information economics is that the combination of high fixed and low marginal costs, high switching costs, and network externalities leads to a dominant firm model. Hence the apparent first-mover advantage, and the race to market.

**The elasticity of the market** The demand curve is rarely linear or static in time and can depend on many other factors, such as the elasticity of markets, advertising expenditure and promotions. So far we have assumed a single supplier. Most markets have multiple suppliers, each producing slightly different goods, but competing for the same customers.

For example, if a number of firms (N) compete in a marketplace, then each may see a market characterized by their own elasticity (Ne). In markets where customers are renowned for their loyalty, such as football merchandise, the elasticity of the market for individual firms will be close to that of the market as a whole. Small price changes will not cause loyal football supporters to switch teams. However, in markets where customers are mainly concerned with price and the goods are universally available, such as standard food supplies, then a small increase in price above a competitor's will result in a large drop in sales.

Some particularly inelastic markets, such as the markets for fuel and cigarettes, can also be affected by tax and other governmental regulations. Such markets need to be very inelastic for people to continue to buy the product despite the high prices, for the tax to be effective. For instance, people have continued to buy petrol despite the recent price increases. As a result, such markets are very sensitive to competitive pricing.

There are a few markets which exhibit negative price elasticity. For example, with high-status goods, if you lower the price, the number of goods sold could fall, as the perceived status of the item will be lost, and the market will consider them too cheap. This example is typical of demand-modified supply-side pricing.

Pricing strategies include pricing:

- for survival: pricing according to how much it takes to keep the company running
- to maximize current profit
- to maximize current revenue
- to maximize sales growth: long-term profits
- to maximize skimming
- quality perception
- land-grab variety, to buy market share

## Testing the market

Economic models, while helping to understand some of the underlying forces, are not the whole story. The real world is often more complex. The economic models are, of course, simplistic and it can be very hard to determine the price sensitivity of a market, especially for a new product or service.

‘WITH HIGH-STATUS GOODS, IF YOU LOWER THE PRICE, THE NUMBER OF GOODS SOLD COULD FALL’

One method that can be used to help you determine how to set your prices is test marketing, although this is not without its dangers. Amazon, for example, caused a furore when they tried to discover demand price sensitivity through market testing, by charging apparently random prices for DVDs. They ended up having to refund customers who felt that they had paid too much, having subsequently seen the lower prices, even though they were willing to make the purchase at the higher price at the time. Surveys and focus groups are other methods which can be used to help determine price sensitivity, but the questions need to phrased with care, as asking people if they would rather pay more or less for a product usually results in an obvious answer! It can be quite hard getting accurate valuations for individual features. The usual approach is to ask people to value the product with differing combinations of features, say, with features A and B, or A and C, or B and C, and then work back to deduce the value of each individual feature.

## What can people afford?

Other demand-side pricing models consider what people can afford. For example, purchase of luxury goods depends strongly on people's disposable income. Wealth is not evenly distributed across society, and typically 20 per cent of the people own 80 per cent of the wealth. Pricing below what the target market can afford may be wasting an opportunity, and may make the market suspicious, as they may think that the product is too good to be true. It is important to target the price at the target market, to fit in with what they can afford, and their aspirations. Strategies like generous trade-in allowances, or deferred payment schemes like hire purchase can help to make desirable items more affordable to an aspiring market.

With high-cost capital items, like cars or personal computers, inelasticity is caused by the fixed rate of customer demand. For example, the average person only changes their car every five years or so, and their computer every three years. If you have just bought such an item, you are unlikely to buy another immediately, whatever the price.

## Skimming the market

Another pricing strategy is to maximize profits from early sales, by what is called 'skimming the market', which is often used in conjunction with a limited production capacity. This involves introducing a new product at a high price, in order to skim off the early adopters and must-haves, on the basis that they will buy the product whatever the price. As the market grows, you need to lower the price to make it more appealing to the majority, but your profits should still be healthy as the price of production will also decrease.

It is better to start too high than too low. Price perception depends on the history of the price. Once a price has been lowered, either by enterprise or by a rival company, it is very hard to raise it again.

❛IT IS BETTER TO START TOO HIGH THAN TOO LOW❜

## Competition-based costing

This leads to another pricing consideration, which is whether to price at, or a little below, the going rate, where it can be determined in a mature market. Of course this will not apply to radically new goods, as a going rate will not exist! If you are part of an extremely competitive market, consideration must also be given to whether you really want to undercut your competitors, as this could trigger a price war, and force tumbling prices where the only winners are the customers.

## Internet pricing

The internet allows novel pricing strategies, since prices can be altered quickly and on an individual basis. Pure demand pricing can be seen, for example, in some internet auctions and pseudo auctions, where the customers set their own price for the goods, perhaps with some minimum reserve. The concept of shareware is also similar, where users send in a contribution for software they appreciate.

## Time constraints

For some goods, the price can vary dramatically with time. Theatre or airline tickets, for example, are worthless after the show has been performed or the aeroplane has left. The minimum price they can be sold at is the marginal cost of filling that seat, which for an empty theatre seat is practically nothing, and for the airline a small amount of fuel. This has led to the development of time-based pricing for internet sales, such as lastminute.com.

## More market characteristics

Apart from the pricing sensitivity, other characteristics of the market need to be defined, such as:

### Size

A fundamental issue is how big the market is, that is, how many products can be expected to be sold at any particular price?

One way to try and ascertain the size of the market is to look at the demographics of the target population. For example, if you are selling a computer game aimed at 10–15-year-old males, then you would need to find out how many boys there are within the sales territory who own, or have access to, a suitable computer. You might even be able to find out what percentage of them like playing computer games and how many could afford your game. This research will provide you with an idea of the upper limit on the market size. With good advertising and distribution you might be able to persuade ten per cent of them to buy the game, although ten per cent is quite generous. For non-essential products, market penetrations of one – two per cent are probably more realistic.

### Feature set

What are they key characteristics of the product? A good way to find out is to ask potential customers for their opinions, such as through focus groups. Having a prototype, or better still, a range of prototypes to show your potential customers can really help here. Even crude mock-ups can help to get your ideas across, and are much more effective than just words and pictures.

While the technical and design side of the company can deliver a range of features, with associated cost implications, the customers have a very different set of priorities, and could value each feature differently. Ideally, the features that you include should be selected on the basis of what is valued by the customer, rather than by the cost of provision. Discard any features that do not pay their way.

It is worth noting that there is a difference between the features that influence a customer's buying decision and those that are actually used. For instance, one of the reasons that I bought my car is because it has four wheel drive, yet I never engage it. However, I'm still glad it's there, just in case. The wordprocessor software that I am using has features, like multiple columns, that I am unlikely to use, but I'm glad the feature is there. My laptop is vastly over-specified for the simple tasks I use it for, but I certainly ticked off the features when I bought it.

"THERE IS A DIFFERENCE BETWEEN THE FEATURES THAT INFLUENCE A CUSTOMER'S BUYING DECISION AND THOSE THAT ARE ACTUALLY USED"

Finding out the customer's perception of the value of each feature is difficult. The usual method is to use the regression technique mentioned above of asking the customer to price various different bundles of features, and then deriving the price perception of each one.

## Sustainability

Is the market sustainable, or is the product something that you just sell once to each customer? Is your product a one-shot wonder, or is it something that will go on selling? What will you do for an encore? Maybe, like fashion goods, it is something that needs continual re-invention, or like a car, you need to produce a new model every year? Some products, such as light bulbs, just wear out every so often and need replacing. Other products such as computers also need replacing every few years, yet this is due to technological advances making the old version obsolete, rather than a system failure.

Some markets, like televisions, are effectively saturated, with only replacement sales occurring. Virtually every household has a TV, and most more than one. Although there have been attempts at market stimulation through technological change, such as HDTV or flat screen technologies, they have not succeeded.

## Exercise

**one** Perform a FAB analysis for the USPs for your product or service. If you don't have a product or service, pick one, like sliced bread. What do you need to do to maintain and enhance the perceived benefits?

## Further reading

Samuelson, P.A. and Nordhaus, W.D. (2000) *Economics*. 17th edn. McGraw Hill

(*The definitive textbook on economics*.)

Varian, H. (1999) *Intermediate Microeconomics*. 5th edn. W.W. Norton

# 14 Reaching the market

‘Half my advertising is wasted, but I don’t know which half.’

(Variously attributed, including to the Philadelphia retail entrepreneur John Wanamaker, about 1885)

MARKETING NEEDS TO DECIDE HOW THE PRODUCT will reach its target audience, and how the audience is to be informed. The first major decision is what channels to use.

The main choice is between a direct sales or indirect sales channel:

- Direct sales means having a direct relationship with your customers, either via your own sales force, or some form of distance selling, such as mail order or a web page.
- Indirect sales means selling via some intermediary, such as a distributor or a retail chain. There may be several intermediaries, and you may, for example, sell to a master distributor in one country, who sells to local distributors, who in turn sell to retail stores, who then sell to the customers.

However, indirect and direct sales are not exclusive, as even when using an indirect route, key accounts and opinion formers may, and probably should, be handled directly.

## Direct sales

For bespoke goods and services, where each sale is different, a direct sales route is appropriate. Examples include consultancy, software programming and high-value goods.

The enterprise employs its own sales staff that manage the relationships with the customers on an individual basis. If you still use a physical bank, you may find that your bank manager is now called a financial services relationship manager. We look at sales techniques in detail in the next chapter, along with some of the issues of managing a sales force, or in new-speak, 'share with you topics in optimizing the performance of a customer-facing relationship team'.

## Protecting the customer

From 31 October 2000, the Distance Selling Regulations give, with some exceptions, new protection to consumers who shop by phone, mail order, on the internet, or digital TV. These protections include, among others:

- the right to receive clear information about goods and services before deciding to buy
- confirmation of this information in writing
- a cooling-off period of seven working days in which the consumer can withdraw from the contract
- protection from credit card fraud.

### Mail order requirements

Mail order is now generalized to distance selling, for example from a web page. Various governmental regulations such as the Mail Order Protection Scheme and the EU Distance Selling Directive, are applicable to such activities to protect against fraud. Credit card companies are likely to impose even more stringent requirements.

Gone are the days when you could advertise some new gizmo for sale, collect the money people sent in, and use these funds to develop and build the gizmo, and to pay for more advertising to fund the next lot. Such financial misuse lowered the barrier to entry, and several fledgling computer companies, some now household names, were originally financed this way. These days, customers' money must be held in a separate escrow account, and not touched until you have dispatched the goods. Most reputable media companies will not accept advertisements unless you can prove that you have the goods in stock and the fulfilment organization ready to deal with the orders and customer enquiries. This is in part self-protection, as the newspaper, magazine or website does not want to get a bad reputation.

Before you set up a distance selling service, there are a number of issues that will need considering.

"GONE ARE THE DAYS WHEN YOU COULD ADVERTISE SOME NEW GIZMO FOR SALE, COLLECT THE MONEY PEOPLE SENT IN, AND USE THESE FUNDS TO DEVELOP AND BUILD THE GIZMO, AND TO PAY FOR MORE ADVERTISINGTO FUND THE NEXT LOT"

**Product** Is the product suitable for selling over the net? Will it survive shipping? It is amazing what can be shipped long distances now – live oysters for example – but you need to ensure that the packaging will stand the rough handling that it is likely to receive. When the product finally arrives with the customer, will they be able to deal with it, or use it by themselves? Can they install it themselves? With one hand?

**Guarantees** What guarantees do you need to provide? You may need to comply with local safety and other regulations, and in any case, a no-quibble return policy goes some of the way towards the testability requirements that were discussed under ACCTO in Chapter.11

**Stocking** You need to have the stock (and the fulfilment organization) in place to deal with the tornado of orders that your advertising will, hopefully, unleash.

**Support** Your customers will expect to be able to get help from a helpdesk, which needs to be there, ready and waiting. Have you provided enough resource? Products like computers generally generate about four calls to the helpdesk for each new customer.

"PRODUCTS LIKE COMPUTERS GENERALLY GENERATE ABOUT FOUR CALLS TO THE HELPDESK FOR EACH NEW CUSTOMER"

**Key accounts** You will want to handle certain key accounts directly. These might include high-volume customers, special deals for the government and the like, and opinion formers like journalists. You might even free-issue one or two products to friendly members of the press.

**Market communications** Writing material that can sell something off the page, be it paper or electronic, is an art form. Unless you possess the skill, get professional help from an advertising agency specializing in this sort of work. Choose them with care, as they will form an intimate part of your team. Inspect their previous work, and speak to their other clients.

**Monitoring** It's no good just spewing the products out and hoping for the best. Good marketing involves constant monitoring of its effectiveness, and feedback to improve both the advertising, and the product. For websites, it is comparatively easy to collect the data about what people are looking at, but much harder to translate that data into information which is useful to the business. For example, the data from log file analyses may show that a significant number of people abandoned a partially complete order, but it will not show why they did. Was the form or the page too complex? Was the price too high? Maybe the site was too slow and they got bored waiting for the response? You can only find the answers to such questions by actually asking the customer.

However, people's answers cannot always be relied upon, and monitoring can sometimes reveal that people's real behaviour is different from what they say in surveys or even in diaries. This is especially true for online activities, or things like TV watching, where people could have hidden agendas. They may not wish to reveal that they actually spend most of their time accessing adult or other 'specialist' content. Such content accounts for something like 70 per cent of overall web page accesses.

## Personalization and targeting

Marketing messages are likely to be more effective if they are aimed at people who likely to be interested, and if they are made as personal as possible. For example, sending information about new cars to someone who has just purchased one is a waste of time and money.

‘MARKETING TARGETED TOWARDS INDIVIDUALS IS BECOMING INCREASINGLY EASY FOR WEB-BASED COMPANIES’

Marketing targeted towards individuals is becoming increasingly easy for web-based companies, due to the amount of data that can be gathered about a user through log file analyses and purchase details. Companies can use this information to target their products and approach on an individual basis. Techniques which are used by some internet companies with considerable success include affinity marketing, where associated goods are offered to a customer ('How about a tie to go with the shirt you have just purchased'), or collaborative filtering, where the commonality between groups of people is used ('Other people who bought this book have also liked . . .').

## Indirect sales

Using a distribution channel means that you sell to an intermediary, who in turn sells to the customer. Of course, the middleman will take a cut along the way, but will also do some of the work. Effectively, the intermediary becomes your customer, rather than the end user.

A typical distribution chain will share the revenue as follows:

- 40 per cent goes to the manufacturer, who uses it to actually manufacture the product, and for development, materials, overheads, research, and, of course, profit.
- 30 per cent goes to the distributor, which pays for advertising and other marketing, stocking and warehouse costs, and their profit.
- 30 per cent goes to the retailer, which pays for local adverts, the shop front, staff and profit.

This means that material costs, which typically represent 30 per cent of the factory gate price, are actually only about 12 per cent of the retail price; and the manufacturer's profit is less than four per cent of the retail price.

You need to choose your distributors with care as, if they do not perform well, you won't have a business. You need to develop good relationships with them, as they control your cash flow. It has not been unknown for a dominant distributor to negotiate for better terms by withholding payments, or even to take over the supplier entirely. Supermarket chains, for example, have a reputation for imposing severe conditions on their suppliers, but in turn can shift large volumes of product.

Distributors tend to have the upper hand in any relationship, unless the brand is well known or the product unique and distributed via many competing channels. If your distributor is not performing well, it can be a long and difficult process to change them. You may be well advised to start off trading by using direct sales, if the product is suitable, and only follow the indirect sales route once your market is established, and your bargaining position is better.

Supermarkets and other high-street retailers are well aware of the value of their shelf-space. They tend to only take on products from new companies if they can be sure of sales. They may require you, for example, to commit to a major advertising campaign, spending millions of pounds to drive customers into their stores to buy your goods.

‘IF YOUR DISTRIBUTOR IS NOT PERFORMING WELL, IT CAN BE A LONG AND DIFFICULT PROCESS TO CHANGE THEM’

## Distributing overseas

For international distribution, the problem is even more difficult. Often, there is a single top-level distributor or agent in each country or territory, who takes on responsibility for and owns the localization of the product, including translation and local regulatory approval. For many countries, distribution is impossible, either by regulation, or, as in Japan, by custom and practice, unless a locally owned distributor is involved.

Controlling and ensuring the performance of such a distributor is always problematic. For a start, there are language and cultural differences. They are also in a different time-zone, are located far away, and operate under different laws in a different jurisdiction. In order to help guarantee that the distributor performs effectively, it can be useful to make sure that they have sufficient skin in the game, such as a significant investment of their own in the success of the product in their territory; therefore, if they do not perform they will suffer actual loss themselves. You may also want to appoint a local independent monitor to keep an eye on them. Export credit guarantee schemes can help lessen the financial risk, but there is a considerable opportunity loss, to say nothing of time and energy wasted, if it all goes wrong.

When looking for distributors in foreign countries, the government (in the UK the Department of Trade and Industry (**www.tradepartners.gov.uk**)) can help to advise you, put you in touch with other people and companies with experience in the territory, help you to find reliable partners, and even fund outward trade missions and visits to trade fairs. Embassies, both the exporting country's in the target country, and the target country's embassy in the exporting country, can also help to find partners, and help with customs procedures, import/export and other regulatory matters.

## Market communications

Market communications is a politically correct phrase for advertising, but has a much wider definition, including other ways of spreading the news, such as mail-shots, PR, and even websites.

Your marketing message will continuously evolve. Media like websites or newsletters (electronic or traditional) allow you to communicate that evolution to the customers. Put another way, don't print too many leaflets at a time, as the message will go out of date.

Related to this is the idea that you should remain flexible, and if you are not getting your message across one way, then try another. Modify the message (and even the product) based on what the market is telling you.

Advertising can be targeted, for example in a specialist magazine or the trade press. Alternatively it can employ a scattergun approach, as in a national newspaper or TV advertisement. What matters is the cost per response, rather than the absolute cost.

The advent of the web brought with it the possibility that advertising would move from a push model, in which the passive viewer or reader is bombarded with messages, to a pull model, where the active reader requests the information that they require. Clearly, sending information to people who don't want it is a waste of bandwidth, while people who are interested may be happy to wade through a long presentation. Alas, that promise has not matured and, if anything, the web is moving to a push model where media is streamed so that additional advertisements, if nothing else, can be inserted. Since it is easy to discover who is online, advertisements are increasingly being personalized or targeted.

Advertising can be used to establish a brand and a corporate identity, as well as actually to sell something. Establishing a brand is expensive and takes time but, once established, a brand can be used to defend a market niche, and, like Calvin Klein for example, it can be used to enhance the value of a basic product.

“ADVERTISING CAN BE USED TO ESTABLISH A BRAND AND A CORPORATE IDENTITY, AS WELL AS ACTUALLY TO SELL SOMETHING”

Advertising is expensive. Major promotions have media costs in the tens of millions of pounds, and a single page in a national magazine might cost in the order of £10,000, in addition to the cost of preparing the artwork. The preparation of advertising material can be quite a slow process, and magazine copy dates might be up to three months before publication, so Christmas campaigns are photographed in mid-summer.

Whatever your product or type of advertisement, it is wise to test your advert out before distributing it widely. Focus groups are a good way to check that the advert delivers the right message, and delivers it effectively.

Websites need to be advertised in order to drive traffic to them. This can be achieved through conventional paper media advertising and publicity, but also by banner ads on high-traffic websites. Such ads typically cost around £10 pcm, that is per thousand impressions, and may result in a few per cent of click-throughs. Don't forget to register with the search engines as well. Search engines often use metadata to characterize the site, so ensure that the keywords field is filled in. You need to include something like:

```
<meta HTTP-EQUIV="KEYWORDS"
CONTENT="enterprise, money, sex, relationships, religion, music">
```

This content tag will give the search engines something to chew on, and should consist of whatever you think is appropriate to be included in the head section of the HTML of the page on your site. However, some search engines are now more sophisticated, excluding common keywords like 'sex', and checking to see that you actually use the keywords you have chosen on the page. The search engine robot's behaviour can be modified using the metatag 'robot', or by including the file 'robots.txt' in the server root directory. A quick web search will reveal the detailed semantics.

## Public relations (PR)

Public relations can be much more cost effective than advertising, as you don't have to pay for the editorial linage, and also people may be more inclined to believe it, as it is an apparently independent and honest opinion.

One option is to use an external PR agency, some of which are very good, but they are even more expensive than lawyers, and essentially all you are buying is their address book. On the other hand, you can employ your own PR person, and do your own PR in-house. You will soon build up your own set of contacts, and it is always much more impressive for a journalist to be put through to someone in the company who really knows, than to deal with an external agency following a script. It is very important that you have strict guidelines about contact with press within the company, and those that need to deal with journalists should be well briefed in order to get the right messages across. Also note that for publicly quoted companies it is a criminal offence to release news that might affect the share price, for example of a big new contract, other than first though the official market channels and news service.

‘IT IS ALWAYS MUCH MORE IMPRESSIVE FOR A JOURNALIST TO BE PUT THROUGH TO SOMEONE IN THE COMPANY WHO REALLY KNOWS, THAN TO DEAL WITH AN EXTERNAL AGENCY FOLLOWING A SCRIPT’

The media is always hungry for material to fill its acres of blank space. However, journalists are as a rule very lazy. Here are a few tips:

- Keep your press release short – no more than a page.
- Write a good headline and something to grab attention in the first sentence.

- Keep it relevant to the target publication.
- Keep it newsworthy, such as a human interest story. If it bores you, then forget it.
- Send it to the right contact personally, not just the newsroom.
- Include contact details for follow-up.

You need to generate news stories regularly, say once a month, to stay in the public eye. The news has to be real and of public interest. However, if you are in the news too often, or if the story is too artificial or trivial, then the journalists and the public will switch off, and like the boy who cried wolf too often, when there is real news you will be ignored. Things that may be important internally to the company, such as the stationery department receiving a new supply of paper-clips, won't make the news. The support division raising money for charity by trying to make the world's longest paper-clip chain might merit a mention and a photo, at least in the local paper and trade press. Bad news travels faster than good, and the plant shutting down because of a shortage of paper-clips may even make the headlines.

## Direct mail

Direct mail is the rifle shot rather than the shotgun of advertising. It attempts to deliver the message directly to those whom you have identified as being interested in it. It is comparatively expensive, and by the time you have printed the material, addressed the envelope and paid for the postage, maybe with a postage-paid return coupon, you might be looking at something like 50p per address. Even with a well-selected list, don't expect more than a 1.5–2 per cent response rate.

Frighteningly detailed targeted lists can be obtained from list brokers. It is amazing how much these people know about you, and they can provide lists of very specific types of people such as homeowners in socio-economic classes A and B, who own a computer and a car, and have a dog. The list broker will charge about 10p per name and address, and for a bit more they will also take all the hard work of envelope stuffing and addressing off your hands.

**Control and record keeping** Direct mail needs to be carefully administered and controlled to keep your customers happy and to ensure that you are not wasting your money. It is bad form to send the same thing to someone twice, or send a promotional flyer to someone who has just bought. If someone phones the company helpdesk, then it's only polite that the person who takes the call should be able to find out what has been sent to them from their record. There are various postal and telephone protection schemes where people can request to be removed from bulk mailings, and in many jurisdictions it is illegal to send unsolicited mail or to make unsolicited calls to people who have asked to be excluded. Your list broker should have removed such names from your list, but it's wise to check, and ensure you are indemnified.

In order to keep track of the response rates of each mailing, you could vary the return address or include a keyword or code on the response forms.

Before expensively generating all these leads, you need to ensure that you have an effective way of dealing with the responses. You will need your strategy in place before you send out the mail-shot, not after, or else you will have mail-bags piled up around the place, filled with rapidly cooling response coupons. If you can't handle the response, it is better not to send out the advertising in the first place. Not responding to the replies will lose you credibility in the marketplace, and with your investors, some of whom will have clipped the coupons just to see what happens.

❛ BEFORE EXPENSIVELY GENERATING ALL THESE LEADS, YOU NEED TO ENSURE THAT YOU HAVE AN EFFECTIVE WAY OF DEALING WITH THE RESPONSES ❜

**SPAM** Do not even think about bulk unsolicited e-mail. SPAM (the electronic sort, not the canned meat) is a *bad thing*. Don't go there, and don't be taken in by the hype of those who want to sell you their CDs of outdated e-mail addresses. SPAM is mildly illegal, anti-social, and most of all, doesn't work.

Don't reply to SPAM, even to request removal from the list; it only validates your address, which can then be sold on to other spammers.

It is much better to build up your own mailing list and community of interest, by, for example, having people register their interest on your website. People are much more likely to do this if you have a strong privacy policy, and never, ever, divulge their e-mail address or details.

## Exhibitions

One thing your company is likely to be tempted to do quite early in its life is to make an exhibition of itself and its new products, especially as governmental subsidy can be had (**www.tradepartners.gov.uk/outward_missions/ outwardmissions/about/index.html**) for overseas exhibitions. Be warned, these are not just jolly outings.

Exhibitions are expensive and hard work, but can be worthwhile once or twice a year, even if only to socialize with other people in the industry. Exhibitions work best as a place for pre-arranged meetings, not cold sales. You can be sure that most of the interesting people in the industry are likely to pass through one of the larger exhibitions, so it's a great time to meet people from overseas.

‘EXHIBITIONS WORK BEST AS A PLACE FOR PRE-ARRANGED MEETINGS, NOT COLD SALES’

Each industry has its own gatherings, and they will go in and out of fashion. For computers and electronics the big exhibitions are Cebit, held in the spring in Hanover, Germany, and Comdex, held in Las Vegas in the autumn. Cebit is the bigger of the two, and is so big that you cannot physically walk round it in the time that it is open. That's big. It is reckoned that each year there are over 750,000 visitors, and 8,000 companies from 60 countries covering some 500,000 $m^2$ of display space. Hotels are booked up years ahead, even in towns hundreds of miles away.

In such a large and crowded marketplace, it's hard to stand out from the rest. Make sure that what your company does is obvious to someone walking by your stand in a hurry from ten paces away. One display looks much like another under such conditions, but a poster saying what you do in five or fewer large words or fewer can really help.

## Preparation

There is a lot to do before actually getting to the show, and you need to set up a team responsible for getting it all together in good time.

**The stand** The exhibitors will just provide you with space, and maybe a blank shell. It's up to you to do the rest. However exhibition halls are often heavily unionized and operate other monopolies, so you may need to hire their equipment and furniture, rather than provide your own. You can hire complete stands, and this is often a cost-effective thing to do, although it may not help you to stand out from the rest.

Your stand design should allow spaces for conducting conversations sitting down, and a place for storage of coats, boxes, leaflets, etc. You should consider catering provision, although a nominated sub-contractor at the exhibition hall will probably provide this. Services such as communications, telephones, power supplies, lights and plants can also be hired from the exhibition organizers, but they will need to be ordered, and probably paid for, well in advance of the event.

Alternatively, you can take a suite at a nearby hotel, and arrange to meet your guests there. You might want to do this in addition to having a stand on the exhibition floor, to provide a location for conversations in a quieter and more private environment. Some companies use trade shows as an opportunity to throw a party for their friends in the industry, such as their actual and potential customers and their suppliers, and also to say thanks to the staff who laboured to get them where they are.

‘ALTERNATIVELY, YOU CAN TAKE A SUITE AT A NEARBY HOTEL, AND ARRANGE TO MEET YOUR GUESTS THERE’

**Exhibits** You have to decide what to display, given the space available and timescale. Will that prototype be ready on time? You will need a good fall-back plan just in case it isn't. What are you going to say about it? You need to get the signage done, and you will need ample supplies of leaflets, and give-aways if you have them. The wording in any literature may need to be checked by legal advisers, and co-ordinated with the PR people, and you will need to allow time for this process. If you plan to use actors and models they will need booking, scripts will need writing, and rehearsal sessions will need to be organized.

You need a co-ordinated way of handling enquiries and contacts, such as a report form, or even an online database. You need booking sheets for meeting rooms, and a staff roster. If you are getting staff uniforms for the occasion, you need to get everyone's measurements, and get them fitted, as well as getting the tee-shirts and baseball caps printed.

The list goes on and on. If you have recruited someone who has done it before, it helps a lot. So does an emergency kit containing a lot of gaffer tape, scissors, blue-tack, pins, corkscrew and bottle opener. Foreign countries have funny plugs and telephone sockets, so you will need converters, and probably extension leads.

Security can be an issue, and anything portable or not firmly bolted down is apt to wander, and even then some of the more desirable things like fancy monitors can disappear. Security mark everything, and secure small items. Make sure there is a lockable store, and collect small items like keyboards and mice at the end of every day.

**Manual** It is very important that everyone staffing the stand gives out the same message and is familiar with everything that is on show, or at least knows how to find someone who is. It's good practice to co-ordinate all the relevant information about a show in a manual, which can be given out to all the staff that will be attending the show, and even to those that are not so that they know what is going on. Your manual can include a pocket for exhibition badges, passes, travel tickets, etc.

If you can get everyone together for a training session beforehand, so much the better. There are some excellent training videos available about what to do, and what not to do.

**Travel and accommodation** Book early. It is always a nightmare. A big exhibition puts a massive strain on the local infrastructure.

## Be attentive

Exhibition time is high-stress for all concerned. It is very tiring, and hard on the feet – wear old and comfortable shoes. However, your company is on show, and you will present a better image and be more successful if the stand staff are smart and attentive, rather than slouching against the furniture. Visitors are more likely to want to talk to you if you are friendly, relaxed and attentive, and make eye contact. Exhibitions can be long and boring, and people like to wander off to see the other stands, but if you are not there, visitors cannot talk to you.

Manning an exhibition stand is like any other sales opportunity, and we discuss the sales process in more detail in the next chapter. The main message is to listen to the customer. The customer is spending time with you because they have a problem that they hope you might be able to solve, unless they are just after the give-away. Listen to what that problem is, take their details or card, and see if you can follow up after the show with a solution.

‘THE MAIN MESSAGE IS TO LISTEN TO THE CUSTOMER’

## Follow-up

After the show, when you get back home, thoroughly exhausted, don't put your feet up straightaway as the job is not yet finished. Hold a post-mortem meeting while ideas are still fresh in everyone's minds, and discuss what went well and what did not, and what you could do better next time. Write it all down, together with what you did, and make sure that you file it somewhere safe, along with the show manual, so that the poor person who has to do it all next year can learn from your experiences. Indeed, you will probably have to start planning next year's show straightaway, as the exhibition space for the next show is often auctioned at the current one, as are the hotel bookings.

You have expensively generated all these leads, and business cards with notes scribbled on them. Are you going to throw them in the bin? You might as well, if you leave responding to them for too long. Sort them in rough order of priority and assign them to the right sales people to follow up. A brief message, for instance ‘Thanks for coming to see us at the show, and we will be in touch shortly. If your need is urgent please call ...' might remind them who you are, and give your address if they have somehow lost the leaflets they picked up at the show.

## Exercise

**one** Analyze half a dozen adverts, preferably for fashion goods or cosmetics. What are the real messages about the benefits of their products that the advertisers are trying to convey? Look at such things as the implied lifestyle and surroundings of the people in the advert. How do these messages differ from the surface message about the product features?

## Further reading

Vyakarnam, S. and Leppard, J. (1995), *A marketing action plan for the growing business*. Kogan Page

(Focusing on the marketing decisions which must be made by smaller companies, this study explores a variety of decision-making processes from start to finish. The authors aim to target the specific needs of any expanding business.)

Winkler, J. (1989) *Winning Sales and Marketing Tactics*. Butterworth Heinemann (Indispensable.)

# 15 How to sell

‘I realized that selling was the greatest career a man could want.’

(Arthur Miller, *Death of a Salesman*, 1949)

SELLING IS QUITE DIFFERENT FROM MARKETING. Selling is customer-facing, and about managing the customer relationship. It is mostly about listening to the customer.

## Understanding the customer

In order to manage a customer relationship effectively, you need to understand your customer. The customer has:

- *Needs*.The customer has a problem that they want to solve, or have a need that they think you can satisfy, which is why they are talking to you. They need to match their needs with the benefits that your solution can provide. Remember FAB from Chapter 1. The main difficulty with such relationships is often one of language, as the customer can only talk about their needs in their terms and their universe, while many technical people have difficulty in explaining their solution in layman speak. It is your duty to jump the gap and to try and understand the world from the customer's point of view. They may have a hidden agenda as well, perhaps to do with internal status and office politics, or just something they don't want to reveal. You need to understand that as well.
- *Concerns*. The customer may have concerns about your solution. They might worry that your company is too small, that there is too much development to do in the time available, or that you are too dependent on a particular critical resource. You need to discover their concerns, and find ways to put their mind at rest.

- *Their place in their organization*. Does your contact actually have purchasing authority? How much can they spend? Very often, your contact may be just part of a larger process, and make a recommendation to someone further up the food chain. If they cannot make the decision, you need to discover who can, and what the process actually is. Has a budget actually been assigned, and if so how much? If not, what is the process for assigning a budget? Again, there may be hidden agendas, and the real process and influence are often different from the official, formal process, with different camps and spheres of influence internal to the company (or even family) in play for any big decision.

To sell successfully you will need to address the customer's needs and concerns, and ensure that they have the authority and ability to make the purchase.

## Stages in selling

Selling occurs not just in moving product, but in all walks of life and forms of interaction. You need to sell your company and business plan to the investors, your budget and new project to the rest of the Board, the project plan to your team, yourself at a job interview and so on.

❛SELLING OCCURS NOT JUST IN MOVING PRODUCT, BUT IN ALL WALKS OF LIFE AND FORMS OF INTERACTION❜

Selling is a process, just like anything else. It has the following stages:

1. Prospecting
2. Pre-approach
3. Approach
4. Survey
5. Proposal
6. Demonstration
7. Close
8. Service.

You need to work through these stages for any selling process. We examine each of them in turn.

## Prospecting

Prospecting is locating the most likely buyers.

We can divide this into:

- cold calling, where there is no existing relationship, and the customer has not previously requested contact
- qualified prospects, where the customer has requested contact, or there is an existing relationship.

**Cold calling** There are many sources that you can use to find the names of companies and contacts to approach. The best method is to get existing customers to introduce you, but if not, look in places like:

- Directors' guides, industry yearbooks.
- Local council offices, and chambers of commerce for local contacts.
- Institutional meetings. Most industries have some form of institutional body, as well as standards committees. These are good meeting places, and the council members and sponsoring companies can give indications as to who the important people in the industry are.
- Web searches and yellow pages.

Make a list, and sort the names in order of priority. The priority can be by likelihood of success, but since that is hard to estimate, you can start with the ones that are easiest to reach.

**Qualified prospects** Qualified prospects can come from:

- Marketing response, from advertising, PR, mailings, etc.
- Existing customers: referrals or additional requirements.
- Service organization. Your service or maintenance organization, and services such as your helpdesk, will have contact with your customers and can provide a good source of information for discovering new requirements from existing customers.
- Exhibitions.

Again, make a list, in priority order. The advantage of qualified prospects is that you know something about them and, since they have asked to see you, there is a stronger prospect of a sale than for a cold call.

**How many?** Realistically, a salesperson can make a maximum of two calls per day, less if there is a lot of travelling to do. That is 400 calls per annum. Maybe five per cent of those calls result in a sale, but probably less, as it may take several calls to make a single sale. Thus, a single person is doing well if they generate 20 sales a year, and four or five might be more realistic. The cost of a sales person in salary, commissions, expenses and overheads for OTE (on-target-earnings) is something like £100K pa. Given that this cost should be equivalent to about ten per cent of the sales they generate, this gives a typical sales figure per person of £1 million, or, assuming 20 sales per year, £50k per sale.

Thus we can see that it is only worthwhile using a direct sales force for high-value sales.

## Pre-approach

Having selected a target prospect, you now need to find out everything you can about them, and prepare yourself before going into battle. Nowadays you can get a lot of information from the internet, and even download things like company annual reports. You should at least have the courtesy to find out about the company's principal products and the names of the senior staff before you visit them.

Remember, your main objective is to find out about the firm's management structure, to check whether you are talking to the right decision makers, and what their key needs and concerns are. If you can find out beforehand why they might want to talk to you, or why they should be talking to you, then so much the better.

**Preparation** You need to prepare beforehand and to take with you:

- a presentation about your company
- visiting cards and brochures.

## Approach

Keep the first meeting entirely social. Remember, you are building contacts. Listen carefully and establish mutual ground, but don't sell. Talk in generalities. If you can move to the next stage easily, then do, but don't push it. You can always come back another time, now that you have established contact.

Sometimes important people in big companies can seem daunting, but there's no need to be afraid as they are only human. As my old salesman said, 'It doesn't matter who they are, they still have seven pounds of hot manure inside them', except that he used a shorter word than manure. Such companies need your new ideas, and if you ring up to make an appointment with the managing director or head honcho, and state your business clearly and honestly, you are likely to get a fair hearing, if not with the head person themselves, then at least with someone close to them.

'SOMETIMES IMPORTANT PEOPLE IN BIG COMPANIES CAN SEEM DAUNTING, BUT THERE'S NO NEED TO BE AFRAID AS THEY ARE ONLY HUMAN'

## Survey

The next stage is to complete the survey of the new contact's needs and constraints, and see where there is a fit. Again, it's mostly a matter of listening.

**Needs, requirements** What are the needs and requirements of the company that you may be able to satisfy?

**Constraints** What constraints are they labouring under? These might be limitations of performance, timescale or budget, or the equipment that they use may have to meet some particular set of standards. They may already have preferred suppliers, with whom they have dealt for years. Another constraint could be their disposition, as they could be worried about buying from a new company, or have other intangible concerns. There may also be internal politics that could affect the acquisition, for example, if the company buys your product, someone else's pet scheme might be jeopardized.

**Budget** What is the budget for the acquisition? Has it been decided yet, and, if not, what is the process of allocating the money? Who is the budget holder who can sign the cheque?

**Structure, contacts, decision points** What is the organizational and management structure? Who are the key decision makers, and what is the process by which a decision is arrived at? Who influences it? Often there are lots of people who can say no, but only one who can say yes. You need to find this person, and talk to them.

**Timescale** What is the timescale for the acquisition, and for the decision process?

## Proposal

Having gathered your information, you can now prepare a sales proposal. You need to get this right, as you only really get one shot at it.

**Sell the benefits to the customer** Remember FAB (features, advantages, benefits) from Chapter 1? The customer needs to understand the benefits for themselves. Some of these benefits may be intangible, or part of some hidden agenda, to which you must tactfully allude.

Stress the USPs (unique selling points) of your offering. What can you do that no one else can?

Price is not normally a big issue, but value certainly will be. You can justify a higher price by a lower lifetime cost, and by increased service, reliability and reputation.

❛ PRICE IS NOT NORMALLY A BIG ISSUE, BUT VALUE CERTAINLY WILL BE ❜

**Presentation and structure** The proposal should be nicely bound in company stationery, with the main part maybe ten pages long with a one-page introduction and executive summary. All the boring detail, fine print and company brochures should be banished to supporting appendices. The proposal can usefully be submitted in electronic format, such as a Word document, but a paper copy, or even better several copies, should always be provided as the definitive document.

A typical sales proposal has the following structure:

- *Introduction*. Sets out the scope and the key players, and summarizes the pitch.
- *Objectives*. Restates your understanding of the customer's needs and requirements. Shows that you have listened and know what you are talking about.
- *Recommendations*. Introduces your solution to their problems.
- *Benefits*. Stresses the benefits of adopting your solution.
- *Financial justification*. Gives a summary of the numbers, and shows the payback period. Justifies the investment in your solutions.
- *Price and conditions*. Appendix.
- *Warranty and service*. Appendix.
- *Outline project plan and timescale,* if applicable. Appendix.
- *Company background*. Appendix.

**Follow-up** When you deliver the proposal, you should arrange for a follow-up meeting to make sure that the contact can understand the document and also to provide yourself with an opportunity to clarify any outstanding points. Don't push at this meeting, as it should just be informational.

## Demonstration

The next stage is to actually demonstrate your system. This is always a fraught time.

**Objectives** Be clear about what the objectives of the demonstration are. At one level, it is to show that system A does tests X, Y and Z. At another, it is part of the validation and due diligence process to show that you are telling the truth, and thus are not likely to have been making up stories about the things that are harder to measure. It is also an opportunity for your people to build relationships with their opposite numbers at the potential customer's company.

**Administration** Make sure that the administration goes smoothly:

- *Who, where, maps, car parking, accommodation*. Make sure that you circulate attendance lists well before the meeting, to guarantee that everyone who should be there knows about it, and knows what their functions are. You will also need to provide the customer with directions if they have not been to your offices before, and possibly arrange transport from the airport or train station, and accommodation for long-distance visitors.
- *Greeting, seating*. Sort out the greeting and seating; make sure that a suitable room is available, and that reception have been informed. Some companies have notice boards welcoming visitors. Make sure you know where the toilets are, if you are meeting in a strange place.
- *Catering: coffee, lunch or sandwiches*. Make sure that the catering arrangements are in place. Nowadays, alas, big business lunches are out of fashion, but sandwiches or other light refreshments are normal if you expect people to work over meal-times. Customs vary from country to country, especially for things like whether alcohol is served. Be sensitive to your visitors' needs, but do what you would normally do.

**Script** It is vital to have a script for the demonstration, and to stick to it. You are probably demonstrating some flaky prototype, and if you don't stick precisely to the script, anything can happen. You need to rehearse as well, just to make sure that it all works, and prohibit the engineers from changing anything between the rehearsal and the actual demo.

Brief everyone who might come into contact with the visitors about the points to emphasize, and the points to avoid. Work out an agreed line for the trickier questions that you can think of, but don't make it all look too slick, or trouble free, or they won't believe you.

**Sum up and agree, then follow up** At the end of the meeting, sum up the demonstration to reinforce the message. Explain what has been demonstrated, summarize the benefits to the potential customer, and arrange a follow-up meeting.

## Close

The sale process now moves into the closing phase. The essence of this phase is to find out what is stopping the customer from signing on the dotted line, and remove the objections, one by one. For example, if they are worried about whether you can actually do the job, offer to post a performance bond – deposit money with a bank, or buy insurance that is paid to the customer if you fail to deliver on time. If they are worried about ongoing maintenance, offer to put the source code in escrow, if suitable, or to give the code to them under non-disclosure. If the financial terms are a problem, maybe a leasing deal would help? Take one small step at a time, stressing the advantages of your solution, until it is a no-brainer to make the deal.

‘FIND OUT WHAT IS STOPPING THE CUSTOMER FROM SIGNING ON THE DOTTED LINE, AND REMOVE THE OBJECTIONS, ONE BY ONE’

Be aware that some people are better at closing deals than others. It is important to be empathetic towards the customer, to be able to easily establish good personal relationships, and to get a good feel for the client's concerns and how to get round them. Some people may have personality traits that make them particularly bad at closing deals. For instance, they may hate to finish anything, and closing this deal, in which they have invested much of themselves, means that they will be somehow empty, and have to find the next thing to do.

## Customers' concerns

Listen to your customer's concerns, There may be some hidden agenda at work that you need to find out about. For example, they may not want to tell you that they have concerns about your financial resources, or they may have a not-quite-so-good deal from an existing supplier, and don't know how to terminate the existing deal. You need to find out about and overcome these objections.

## Kicker

Finally, in some cases, if the client is wavering on the edge, you can offer a kicker, for example a limited period discount offer. Kickers are a bit like ultimatums, and cut both ways. You need a fall-back plan in case the bait is not taken, and you need a way to get back to normal relations.

## Service

When the deal is done, the work on the relationship continues, but its nature changes to that of management of the relationship. The sales team becomes the main contact and conduit of information between the company and the customer. It is very likely that not all of the information will be good news, and the sales team may have to pass on bad news such as missed milestones, and then manage the resulting disturbance.

**Communications** The primary mission of relationship management exercises is to stimulate communication, both formal and informal, such as social events. The customer is not the enemy – they are the ones paying for the work, and ultimately everyone's salaries.

**Contact point** The sales team acts as the primary point of contact. They act as a pseudo-customer, internal to the company for the benefit of the development team.

**Regular liaison** As part of the communication, the sales team need to hold regular liaison meetings with the customer. These can be planned to conveniently correspond with the external milestones of the project.

**Early warnings** Things do not always go smoothly, and the earlier that the customer can be warned of trouble, the easier it is to deal with. Disturbance and expectation management are a key activity.

**Specification changes and the consequences** Customers will inevitably change their minds. Hopefully, there will be a change control procedure in place, which needs to be understood by all. The sales team has an important role to play in ensuring that the procedure is adhered to, rather than, for example, the customer ringing a junior member of the development staff direct. Change is not usually without consequences, such as extra cost or delay to the project, and one of the more difficult tasks of relationship management is to get the customer to understand and accept these consequences.

## Planning and records

You need to keep consistent records of the sales process, just in case you fall under a bus and someone else has to take over. For a small company these can be manual, but are more likely to be electronic these days. You can get several excellent packages to do this, for example systems that run under Lotus Notes. Better ones are now network aware, and therefore can be securely accessed from the field, from a cell-phone or a PDA, or via the web. Key documents are the prospect list and sales forecasts.

### Graded prospect list

The prospect list contains key facts about each prospect, and grades them by the likelihood of them becoming a customer. The prospect list also records actions done and to do, and lists all contacts with the potential customer. In electronic formats you can mine down for more information. A typical format might be as shown in the illustration on page 326.

### Sales forecasts

The sales forecast takes the information from the prospect list, and presents it as the expectation of what sales can be made according to your prospects, by time period, and by product. A typical format is:

*Prospect List for ..........................Sales Group*

| | | | | | | | | *Previous contact* | | | *Planned next contact* | | |
|---|---|---|---|---|---|---|---|---|---|---|---|---|---|
| *Name* | *Address* | *Phone* | *Fax* | *Contact's name* | *Decision makers* | *Potential value* | *% of success* | *Date* | *Who* | *What action* | *Date* | *Who* | *Actions planned* |
| | | | | | | | | | | | | | |
| | | | | | | | | | | | | | |
| | | | | | | | | | | | | | |
| | | | | | | | | | | | | | |

| Sales Forecast for the Widget Division | | | | | | | | | |
|---|---|---|---|---|---|---|---|---|---|
| Sales account | Total amount | ------Time analysis--- | | | | ----Product analysis------ | | | Comments |
| | | Q1 | Q2 | Q3 | Q4 | Nuts | Bolts | Screws | |
| | | | | | | | | | |
| | | | | | | | | | |
| | | | | | | | | | |

where nuts, bolts and screws represent three different products.

## Measurement, control and commissions

Sales organizations are quite difficult to control. To control, you must first measure.

> TO CONTROL, YOU MUST FIRST MEASURE

### Measurement

**Call analysis** Call analysis measures the effectiveness of the sales force. What percentage of calls result in a sale? How do your sales figures compare to the industry average, and how do they vary with time? What factors contribute to your success rate? Does the sales force need better quality leads, or more support?

**Sales cost analysis** Sales cost analysis looks at the overall sales process. Measures include:

- cost per sale
- response rate to advertisements and other market communications
- timeliness – how long does an enquiry take to be answered?

**Individual measures, targets** Individual measures and sales targets for each member of the team can be set. Since commissions are often linked to these targets, they are not without controversy.

## Control

There are traditionally three ways of organizing a sales team. Each has advantages and disadvantages.

**By product** If the team is organized by product, then the members of the sales team are product specialists, with a deep knowledge about what they are selling. The expert nature of the salesperson is a particular advantage of this model, as they are able to speak with authority about their specialist area, and help solve customers' problems. The disadvantages of this model is that they have to do a lot of travelling, and may be blind to other product opportunities which they might come across.

**Geographical** If the team is organized by territory, the salesperson can build relationships with the potential customers in their territory, and can quickly respond to their needs, without having to travel too far. However, the disadvantages of this model are that the salesperson is a jack-of-all-trades, and must be able to sell all the different products that the company produces.

**By channel, industry sector or key account** This is a version of the geographical territory method, and has similar advantages and disadvantages. Industry sector specialists can work well, where this is appropriate.

‘SMALL COMPANIES DON'T GET MUCH CHOICE ABOUT HOW TO ORGANIZE THEIR SALES TEAMS’

Small companies don't get much choice about how to organize their sales teams, but they will probably not have many products. Small companies will have a small sales team that covers everything. For larger companies, one method of organization is to have geographical or industry-based territory sales staff at the sharp end, and then have flying product specialists in support, who can be called in to do the technical sales.

## Commission

Sales staff are traditionally rewarded and motivated by commission, and other rewards such as a fancy car on loan to the salesperson of the month. These rewards are on top of a basic salary, which allow the sales team to keep body and soul together.

Don't stint the commission. Sales are the lifeblood that keeps your company going, and the sales team earn their ten per cent. It's no use complaining if they are the best-paid people in the company, because without them at the front end, nothing else would happen. On the other hand, only pay out the commission when the customer pays the company, otherwise you will get a lot of cancelled orders.

## Exercises

**one** Go and find a friendly salesperson and spend a day on the road with them.

**two** Draw up a prospects list of your first ten customers. What would your 'elevator pitch' to them be?

## Further reading

Fenton, J. (1998) *Close! Close! Close*! Management Books 2000
(Also other books in 'The Profession of Selling' series.)

**www.iops.co.uk/** The Institute of Professional Sales

**www.ismm.co.uk/** The Institute of Sales and Marketing Management

# Growth and exit

# 16 Coping with growth

> "For most of those which were great once are small today; and those which used to be small were great in my own time. Knowing, therefore, that human prosperity never abides long in the same place, I shall pay attention to both alike."
>
> (Herodotus, *The Histories* I. 7, fifth century BC)

OK, YOU'VE ESTABLISHED YOUR COMPANY, BUILT THE product and made some sales. Things are going quite well.

In this chapter we discuss the problem of growth. Growth brings its own problems. Starting a company is very different from running it. As the company grows and matures, the skills required will change. However, growth is unavoidable, and a company needs to grow in order to survive. There are very few companies who manage to stay still, especially as the world around them changes.

## New markets

There are only two ways to grow: horizontally or vertically. Horizontally is selling the same things to new customers, vertically is selling new things to the same customers.

Of course, you could decide to start something completely new, selling new stuff to new people, but that would be tantamount to starting a completely new business. The danger with starting a whole new enterprise is that you are liable to take your eye off the ball of your existing business, to the detriment of both. If you really want to start a new enterprise, it would probably be better to sell off your interests in your previous enterprise, and concentrate on the new one.

### Horizontal

Horizontal expansion is selling the same or similar goods and services to new customers. One way in which you might reach a new customer base is by targeting new geographical areas by, for example, opening a new shop in a different area, or by starting to export to a new country. Of course, you can always find new customers in the same area too. To attract new customers from within your existing target market, you may need to try out new approaches such as increasing your marketing spend, lowering the cost of the product, or finding a new use for the product. A classic example of horizontal expansion is the humble, but hugely successful Post-it note, which was developed from an ineffective glue by 3M.

### Vertical

Vertical market expansion is selling new goods or services to the same or similar customers. This can be achieved through the introduction of a new updated model, as, for example, with personal computers or electronic games, or by creating add-ons to go with an existing purchase, such as new channels for cable or satellite TV, or accessories. You might sell other things your customers need, so as to become a one-stop shop for them. For example if you sell paint, you might consider selling paint brushes and brush cleaner, as well.

Another form of vertical expansion is vertical integration up or down the value chain, for example by acquisition of suppliers or distributors. You do not expand the market doing this, but you get to keep more of the money generated by that market.

‘ONE OF THE WORST PROBLEMS IN THE EARLY GROWTH STAGES OF A YOUNG ENTERPRISE, IS THE SIN OF HUBRIS’

## Problems of growth

One of the worst problems in the early growth stages of a young enterprise, which is often repeated, is the sin of hubris. Because everything is going so well, the management think that they can walk on water, and dramatically over-expand the company, or move into areas that they really know nothing about, or spend too much time being feted at industry fests, with the inevitable collapse.

### Inside the tornado

In Chapter 11, we discussed the chasm between products suitable for early adopters, and products for the majority, according to Geoffrey Moore. However, once the chasm is crossed, there are other problems to solve.

Geoffrey Moore discusses these subsequent challenges in his second book, *Inside the Tornado*.

If your product is a real success and meets the needs of people, then you may have a 'success-disaster' on your hands. A success-disaster is when your manufacturing processes cannot keep up with the demand, creating shortages. Your helpdesk gets swamped with enquiries from angry would-be customers. Your inability to deliver will mean that you develop a bad reputation, and allows your competitors into the market that you pioneered. In an attempt to make up lost ground, the company takes on more commitments than it can deliver, it overtrades, runs out of money, and then collapses – or is ignominiously taken over by an older, slower but richer rival.

It is possible to avoid such disasters, but it needs a lot of discipline. Do not bite off more than you can chew, and make sure that you are in control of your situation. Measures that will help you to control your growth include preparing accurate and timely sales forecasts, and considering carefully the benefits and disadvantages of tackling new markets. Instead of trying to sell to everyone, establish a dominant position in a niche market, such as a geographic sector, before moving on to conquer the next niche. Raise the price to control demand, at a level with which the factory can cope. The price can be gradually lowered as demand weakens. However, a high price will improve profitability, and will mean that the product is perceived as a top-end product, which can be very beneficial.

## Divide and conquer

People naturally work best in small groups. Mankind's evolution from primitive hunter-gatherers has left our brains wired to maintain close relationships with only around 15 people, and to work best in teams of about seven – the optimum number of people in a hunting group.

The hang-over today is that companies tend to have break points at multiples of seven, that is at headcounts of about seven, 50 and 350, where another level of management is required at each transition point. Passing through the transition points can be painful, and can involve a major change of company culture. In a group of seven people you will know everyone, and are pretty much aware of what they are all doing. With 50 people, you will still probably know everyone's name, even if you don't supervise them daily. However, with 350 people you won't be able to put names to faces. You will lose touch with the workforce, and will only talk to their managers.

Putting in an extra layer of management is always hard. It is always difficult to take a close-knit group, and split it into sub-groups, each with its own manager. If you are lucky, there will be a natural plane of fission, and a natural leader-in-waiting for the new group, but that is rarely the case. You will be faced with breaking up a work family, changing the role of the leader of the original group and appointing either an outsider, or promoting one of the existing team members over their colleagues. In such situations, the original leader often feels resentment, because they feel that their responsibility has diminished, and that they have lost control of the team that they have done such a good job with and have put so much work into. You need to explain that, far from diminishing their responsibility, you are increasing it.

‘PUTTING IN AN EXTRA LAYER OF MANAGEMENT IS ALWAYS HARD’

## Managing change

One of the more difficult management tasks is the management of change. Within an organization, change is inevitable. When major change happens, for example a project is cancelled, or a group is split up, people go through four stages of adaption, typical of any loss or grief process. Compare this with your own experience should you have been so unfortunate, or that of a friend, of situations such as the loss of a loved one, being made redundant, being told about a serious medical problem or going through a divorce. Most counselling is based on helping people through these stages. Different people take differing amounts of time to get through these stages, but the general experience is common to all.

They are:

1. *Denial*. Disbelief at the news: ‘This can't be true’ or ‘This cannot be happening to me’. People refuse to accept the change, and would

rather bury their head in the sand, and hope that it goes away. The best way to help someone cope with this stage is through the provision of information, as much and as accurate as you can find. The person needs to know exactly what the change will be, who it affects and how, what timescale it will operate on, and what the alternatives are. They will need help to plan, and time to come to terms with the change.

2 *Anger*. The next stage is anger and blame. 'How dare they do this?!' This will be a tricky time, as people will probably act irrationally, or may just withdraw, feeling unvalued. It's no good telling them to snap out of it. At this stage, they need to be listened to and empathized with – they are not yet ready for solutions but need their responses and reactions acknowledged.

3 *Resignation*. After anger there comes resignation and the beginnings of acceptance. People start to explore their new roles and opportunities. They need help and information about the possibilities open to them, and involvement in planning and goal setting. Focus on short-term wins, as they need to see the benefits of the change.

4 *Acceptance*. The final stage is acceptance of the change and commitment to the future. The person will have a clear sense of where they are going. At this stage, they need encouragement and positive feedback to consolidate the change.

## Communication

Communication can start to break down as the company grows larger. It is often the first thing to go, due to the fact that you no longer see everyone on a day-to-day basis. The game of Chinese whispers shows how messages get transmuted as they are passed on by different people (see crock of shit example in box), and the same thing happens when messages pass through layers of management. The directors of the enterprise may have clear vision, but this may be muddied by the time it reaches the workers at the sharp end. Equally, those workers may have a clear idea of what's wrong or what needs to be done, but their messages may be very diluted by the time they reach those who can do something about it.

‘COMMUNICATION IS OFTEN THE FIRST THING TO GO’

'In the beginning was the plan, and then the specification;

And the plan was without form, and the specification was void.

And darkness was on the faces of the implementors thereof;

And they spake unto their leader, saying:

"It is a crock of shit, and smells as of a sewer."

And the leader took pity on them, and spoke to the project leader:

"It is a crock of excrement, and none may abide the odor thereof."

And the project leader spake unto his section head, saying:

"It is a container of excrement, and it is very strong, such that none may abide it"

The section head then hurried to his department manager, and informed him thus:

"It is a vessel of fertilizer, and none may abide its strength."

The department manager carried these words to his general manager, and spoke unto him saying:

"It containeth that which aideth the growth of plants, and it is very strong."

And so it was that the general manager rejoiced and delivered the good news unto the Vice President.

"It promoteth growth, and it is very powerful."

The Vice President rushed to the President's side, and joyously exclaimed:

"This powerful new software product will promote the growth of the company!"

And the President looked upon the product, and saw that it was very good.'

(From the definition for Snafu principle in the Jargon File version 4.2.3: **www.tuxedo.org/ ~esr/jargon/html/index.html.**)

## Conscious effort

Maintaining good communication requires conscious and continuous effort, and needs to be deeply ingrained into the company culture. Since knowledge is power, especially in office politics, it is all too easy for the middle layers of a company to block the flow of information. If the information consists of bad news, people may not want to admit it to their managers or even to their peers. It may be something that they haven't understood, and don't want to show their ignorance about, and therefore don't pass the message on. An open and supportive company culture can help overcome these failings, and help develop better communications. You should aim to 'share and enjoy' information. This was the motto of the complaints division of the Sirius Cybernetics Corporation (the ultimate gaggle of incompetent suits) in Douglas Adams's *The Hitch Hiker's Guide to the Galaxy*. The irony of using this as a cultural recognition signal appeals to hackers.

Management by wandering about can often help communications. If you spend all day in your office shielded by personal assistants, you won't be able to pick up on how people are feeling, what they are saying, and what general morale is like. Walk around and talk to people. It is important that you are visible to the people doing the work, that you are approachable, and encourage workers to come to you with any concerns. On the other hand, you must be careful not undermine the intermediate layers – if you find a problem, fix it through the formal channels. The role of senior management is leadership, listening and mentoring. Don't make your wanderings a formal tour of inspection, surrounded by flunkeys, as you won't learn anything. Try working for a day on the helpdesk, or whatever else counts as the coal-face for your particular company, anonymously if need be. Even better is to insist that all your senior staff, including yourself, take their place on the shop floor, and do sales calls at regular intervals, such as once a quarter.

## Formal channels

Formal channels of communication, up and down the management tree, are the arteries of the company, and allow information to be disseminated and to get through to where it is needed. Weekly reports, regular management meetings (between a manager and their direct reports), and one-on-one reviews are the stuff that keeps the company's heart pumping. In addition, you will also need a company intranet, project and team websites, mail lists, and company directories, as well as all the documents involved with formal reporting such as project proposals, milestone reports, sales figures, budget requests and approvals, travel requests, staff information and HR requests, etc.

Formal channels of communication work best up and down the tree, but can be quite slow and subject to noise. Therefore you will need informal methods of communication to supplement them.

“FORMAL CHANNELS OF COMMUNICATION WORK BEST UP AND DOWN THE TREE, BUT CAN BE QUITE SLOW AND SUBJECT TO NOISE”

## Charters

Charters are a way of breaking a big company down into a set of smaller co-operating companies. The process involves setting goals and operating parameters for each of the smaller chunks, which not only helps the chunks themselves to understand what they have to do, but, because the charters are published, it also helps others to understand what they do, and how to relate to them. Smaller units can act more like small companies, and will hopefully have gains of efficiency, internal communication and spirit. They also give something to measure performance against, for example for bonus schemes.

## Newsletters and suggestion schemes

Company newsletters and in-house magazines can be a two-edged sword. They can help to propagate the vision and message downwards, but can also highlight how out of touch the senior management is with the workforce. For instance, showing pictures of the chairman on some jolly outing, has just as much potential to generate resentment as it has to inspire.

Initiatives such as suggestion schemes and committees, works councils and quality circles can, in principle, help the communication to flow back up the tree, and might also be useful as a forum in which people can let off steam.

However, such schemes need to be taken seriously, otherwise they can just degenerate into talking shops. Senior members of the company need to demonstrate that they are really committed to making such initiatives work. It is also essential that visible action is taken as a result of their deliberations and recommendations, or else the meetings will get a poor reputation, and soon no one will bother to show up.

### Company meetings and informal events

Informal communications can be greatly enhanced by company meetings, divisional off-site meetings, Christmas parties and other similar events, if structured right. The real purpose of such gatherings is not to listen to the chairperson's words of wisdom (although the chairperson may disagree!), but to get people to mix informally. A company meeting should be held in unthreatening, conducive surroundings to encourage people to talk to other members of the company that they would not normally come across. Such meetings are a time and place for management to actively listen, and for people to make contacts, and ask questions. Why is the mail erratic? Perhaps the people in the post room can tell you. The enterprise is a community, and people come to work for more than just the wages. Of course, such meetings can provide a forum for discussions beyond work, and you may find other people with similar interests or talents to you, and end up forming a recreational group such as a company jazz band or bowling team.

## Formalization and structural change

As the company grows, it will inevitably become more formal. Systems are devised, often for good reason, so that people can keep track of things, and ensure that those who need to be informed get informed. Despite your best efforts, people start wearing suits, and someone will invent a form for ordering paper-clips, and insist on its use. Big companies have culture, but small ones have personalities, usually that of its founder. As the company gets bigger the light of that personality is harder to see in the outer reaches, and systemization sets in.

‘BIG COMPANIES HAVE CULTURE, BUT SMALL ONES HAVE PERSONALITIES’

Big companies are essentially different, slower and more rigid than start-ups, as formal systems are more rigid than the informal varieties. As formal systems set in, the company will gradually ossify, and become slower to react to change or opportunities. Every now and then there may be a shake-up or palace revolution, but the old ways will be discarded only to be replaced by new formalities.

Innovation is the last thing that many large companies need to do. It confuses their market, and, unless their market is saturated, they do better spending their money on exploiting their existing products rather than generating new ones. The only time that large companies need to innovate is when their rivals introduce something new, and even then they can probably get the development done by a specialist outside contractor, and thus not have to disturb the existing culture.

That is why start-ups can sometimes run rings around big companies. Large companies are starting to recognize this, and are starting to encourage and invest in spin-out companies. This involves the company encouraging innovative members of their staff to set up their own company to follow a project in which the main company is not directly interested. The new company, freed of the constraints of the parent, can regain its youth and move more quickly. The investment is a small risk for the parent company, and just might be an insurance policy for the future.

## Different skills, different people

The skills required to run a company are quite different from the skills needed to start one. In a start-up, every day is an adventure with new things happening, whereas running a mature company it is more about risk reduction, and doing the same things in a reproducible manner. The wise entrepreneur will realize this, and will recruit outside skills, such as an experienced CEO, to take over in a timely fashion, before they are forced to do so by their investors. 'Dr XXX, one of the pioneering founders of the company, will be devoting more of their time to the research division' is the usual rubric. It is rare for the two sets of skills to be held by one person, as they involve different emotional outlooks, with one role requiring a risk taker, the other a risk avoider. If you do not have the skill set to run a maturing company, it will be much better for you to move on to the next new thing, and let those who enjoy doing the same thing every day make money for you. Knowing when to do this, and having the modesty to do so, is one of the distinguishing features of great entrepreneurs.

## Cash and overtrading

Growing, especially growing fast, requires wads of cash. New plant, new markets, and new staff start costing money immediately, but do not contribute in terms of bottom-line profitability for some time. This puts a considerable burden on the cash flow, and you will need to find extra cash to fund the growth. Even though the enterprise may be more profitable eventually, there is a period of hiatus when growth takes place, a bit like changing gear in a car. New plant, offices, shops and factories need capital investment from somewhere. Raising the cash is probably the biggest single limitation on the rate of expansion.

Beware of overtrading, which inevitably you will do. Taking on commitments that the company cannot fulfil is a very common hole to fall into. It's hard to turn down potentially profitable work and opportunities as they are offered. If your major client offers you a juicy new contract, but one that means growing the company faster than the resources on hand, it takes a lot of character to turn it down, or even to ask for the money up-front. Turning it down may jeopardize other and existing work from the same source. Taking it on may mean having to raise more money on unfavourable terms, and you may overload your staff, your machinery and your overdraft, none of which work efficiently when near to capacity.

‘BEWARE OF OVERTRADING, WHICH INEVITABLY YOU WILL DO’

## Exercise

**one** Pick a medium-size company with whose products you are familiar. If you can't think of one then use, say, your nearest independent baker. What are their growth prospects? What markets could they move into, and how much would it cost?

## Further reading

Baghai, M., Coley, S. and White, D. (2000) *The Alchemy of Growth*. Texere Publishing

Barrow, C., Brown, R. and Clarke, L. (1995) *The Business Growth Handbook*. Kogan Page

(Contains ideas on business development which have been taught at the Cranfield Business School and disseminated to a large number of firms.)

Hamel, G. and Prahalad, C.K. (1996) *Competing for the Future*. McGraw-Hill Publishing Company

(Provides a model for attaining industry leadership. It explains how to balance the demands of current businesses with the drive toward leadership in the markets of the future.)

Johnson, S. (1997) *Exploring Corporate Strategy: Text and Cases*. Prentice-Hall

(Brings together the underlying concepts, analytical methods, processes of development and problems of corporate strategy, enabling readers to understand the role of corporate strategy within a variety of organizations and providing guidance in the formulation and implementation of strategy.)

McKeran, D. and Flannigan, E. (1996) *Shaping the Entrepreneurial Company*. Management Books 2000 Ltd

(Discusses the importance of the entrepreneur as the driving force behind the creation of the entrepreneurial enterprise, and examines how entrepreneurs develop and build successful companies.)

Muzyka, D., Churchill, N. and Leleux, B. (1998) *Entrepreneurial Management*. McGraw-Hill Publishing Company

(Offers a wide-ranging treatment of entrepreneurial management including: start-up, buyouts and buy-ins, managing rapid growth including venture capital and financing, crisis management, and preserving and instilling entrepreneurship.)

**www.transitions.co.uk/SOQ/soqintro.htm**

(Shai Vyakarnam's strategic options questionnaire.)

# 17

# Valuation

'A cynic is a man who knows the price of everything and the value of nothing'

(Oscar Wilde, *'Lady Windermere's Fan'*, 1892)

## An introduction to a difficult issue

There will be many different things that motivate and drive you to start a new business, such as a passion for your new concept, the excitement, challenge and energy involved and, of course, money! But how much will your new business be worth, and how much money can you expect to make from it?

When considering these questions, there are two important points that you should take into account. First, new ventures that are started purely on the premise of getting rich are much less likely to be successful than those that are driven by the passion for making a new idea or concept succeed. Second, valuation is about much more than just how much your business will raise when you eventually sell it. Valuation includes non-monetary and intangible worth too, such as the enjoyment that your team will get from working together, and the effect of the business on the community.

Knowing what your business is worth will be vital when you start looking for investors, as an investor's motive for putting money into your business will be the share of its worth that they will expect as a return. The value of the company will affect how much money the investor will be prepared to put in and how much of the business they will get in exchange. Investors talk about the value of the business 'pre-money' and 'post-money'. If the value of the business 'pre-money' is, say £2 million, and the investor is investing £1 million, then the 'post-money' value is £3 million, and the investor will start negotiations expecting to own 33 per cent of the business. This illustrates why you need to be able to value the business as soon as you start talking to potential investors.

An investor could receive a regular stream of interest, with very little risk, by putting their money in the bank. By investing in a company, the investor will be hoping for an increased yield (percentage rate of return) on their investment, through a stream of future profits and dividends. Of course, the potential of increased yield also carries a higher risk associated with the future cash flow compared to putting the money in a bank, especially for early-stage companies. As a result of the initial high risk, the valuation of the company's worth will be low to start with, although as the company matures, the risk of not making future profits declines and the value of the company will go up.

‘AN INVESTOR COULD RECEIVE A REGULAR STREAM OF INTEREST, WITH VERY LITTLE RISK, BY PUTTING THEIR MONEY IN THE BANK’

However, if a company's early promises of vast profits seem to be exaggerated and reality suggests that a lower level of profitability is more likely, the company's value is likely to go down. This is one of the reasons for the decline in the value of dotcom companies, as they failed to meet their predicted profitability.

So, the first and one of the most important lessons on valuation, is that:

> 'The value of the business changes over time.'

This may seem obvious, but many new entrepreneurs consider that the worth of their business just refers to the size that they expect their business to be when it has reached maturity, and therefore talk to investors about how they expect the business to be worth, say, '£50 million in five years'. Although this may well be the case in five years' time, and I hope it will be, you can be quite sure that your business will not be worth £50 million today, if all you have is the idea, and possibly a rough prototype. In the early stages of a start-up company, even though you may have a great, wonderful new idea for a business, the actual worth of the company will be very little, in fact probably only a little more than nothing. A business will only have real value when the fantastic new idea has been proved, that is when you have built the prototypes, grown the business, sold real products, and even made a profit, and the risk of the future profit stream can be seen to be low.

How exactly does the value increase in relation to the growth stages of a company? Is it a 'straight-line process' or are there key milestones and steps in value? How can you measure, or at least estimate, the value at certain points during the development of a new business? Are there certain points where it is more advantageous to bring in investors than at other times? These questions are answered later in this chapter. However, before we start looking into values, finance, accounting and such things, we first need to consider the second lesson on valuation:

> 'It is an art more than a science – there is no proven formula that will determine the value of your business.'

Although there are some guidelines and useful tools that can be used for valuing your business, which are covered in this chapter, there is no guaranteed analytical method for setting a value to your business. This concept can be difficult for physicists, engineers and stubborn strong-minded people to grasp, but it is important that you get used to this idea from the start. Since the future is hard to predict, different people will have different perceptions of the risk of future profitability, and hence different values for the company. The more that you can reduce these uncertainties, the more accurately you can value the company.

'THERE IS NO GUARANTEED ANALYTICAL METHOD FOR SETTING A VALUE TO YOUR BUSINESS'

## The concept of value: buying and selling

Because there are no firm rules or laws for calculating the value of a company, and because value can change dramatically according to different circumstances, valuation is a very tricky issue.

There are some circumstances when you can be very sure about the worth of something. For example, a £20 note is worth almost exactly £20 in all circumstances, unless it is converted to another currency such as dollars, euros or gold. If someone started selling £20 notes for £15 each, everyone, including myself, would be queuing up to buy them, as you could be sure that you are getting a bargain, although even then there would be the suspicion that these notes were forged, stolen or otherwise risky. However, consider the value of a second-hand car. If you have ever sold one, you will know that the value of an asset is by no means certain. You may think that the car is worth £2,500 because you have looked at similar cars advertised in the paper, but when you look closely you realize that these cars are being sold by dealers, along with a warranty. Therefore, your car will be worth less, but how much less? You may decide to advertise your car at a price of £2,000 to start with, on the understanding that any buyers will probably offer under the asking price, and you would actually be very happy to take £1,500 for it. However, after the car has been on the market for four weeks with no takers, you will probably be willing to take £1,000 to clear it. Therefore the value of your car could be worth anywhere between £2,500 and £1,000 depending on the circumstances, your attitude and your confidence.

Perceived value depends on whether you are buying or selling. From a seller's perspective, the perceived value of the asset will be high, as they will be thinking about all the good features, characteristics and most of all the future potential. However, the buyer is more likely to think about the risks, problems and the potential downsides of the asset, and hence their perceived value of the asset is likely to be lower. Added to these perceptions is the fact that the buyer and seller are both business people trying to get the best possible deal, so there will inevitably be some degree of gap or range between the buying price and the selling price.

Because there are so many variables involved in these calculations, the process of estimating the valuations for a particular new business is very non-scientific.

## How do businesses grow in value?

As already mentioned in the introduction to this chapter, your business will have very little worth at the beginning of its life when you just have the idea, but will be worth a great deal more when you have introduced products to the marketplace and have paying customers. But exactly how does the value change over time, and what types of events can cause the value to increase?

First, it is important to understand that the value of your company does not just increase with time, and is not necessarily dependent on how long you have been in the business. In order for the value of your company to increase, the risks to future profitability need to be reduced. This can be achieved through measures such as improvements to products, making them more likely to be successful in the marketplace, or developments such as a new investment that will grow the business, and hopefully its revenues and profitability.

“FIRST, IT IS IMPORTANT TO UNDERSTAND THAT THE VALUE OF YOUR COMPANY DOES NOT JUST INCREASE WITH TIME”

There are a large number of activities that will be important in growing a new business venture, but some of these will be much more important than others in growing value. The activities that make the largest contributions to an increase in value of a company are those that demonstrate endorsement or success of your business, for instance:

- Getting initial funding. This shows that someone else believes in you, and is willing to invest their money in the venture.
- Building a team. With more than just one of you working on your idea, it should improve. If you assemble a team, this will constitute more people who are committing to the idea.
- Making a working prototype. Once you have demonstrated that your idea really does work, it is easier to make people believe in it.
- Being able to demonstrate to a customer that the idea is effective. When your prototypes are good enough to impress a customer, you will be near the point of making enough sales to make a profit – this is a clear indication of initial success.
- Having a profit forecast for the year and meeting it.

“ONCE YOU HAVE DEMONSTRATED THAT YOUR IDEA REALLY DOES WORK, IT IS EASIER TO MAKE PEOPLE BELIEVE IN IT”

The last point is probably the most important in terms of building potential customer's confidence in the value of the business. It adds credibility to your future forecasts.

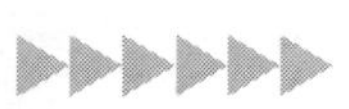

## Growth in value

To illustrate how businesses grow in value, let's consider CEC Design Ltd, and plot a rough timetable and key events. CEC Design (not a real company) has been started by Steve and Dave to make holographic projectors, based on their research at university for their PhD theses.

| *Date* | *Activity* | *Value* | *Importance for the business* |
|---|---|---|---|
| *Year 0* | | | |
| Nov | Initial idea | 0 | Critical for business; not too important for value |
| Dec | Work on the idea to make it into a solid concept | 0 | Important for business; no major increase in value |
| Jan | Initial prototype; preliminary testing | 0 | Important for business; no major increase in value |
| Feb | Decide to set up a company; get Joanna and Louise to join the team; obtain funding from friends and family | 20K | Important for value; important for business |
| *Year 1* | | | |
| Mar | Work on business plan, continue technical development work; start talking to customers about their needs | 20K | Important to business; no major increase in value |
| Jun | Beta prototype works well; first test customer is happy with results; first sale to a paying customer; attract angel funding from a contact of their old professor's | 100K | Important to business; important to value |

| | | | |
|---|---|---|---|
| Sept | Conclude a deal with a major distributor | 500K | Important to business; important to value |
| Oct | Sale of five projectors to multi-national company; start negotiating for first round of VC funding; understanding bank manager agrees extended overdraft | 750K | Important to business; important to value |
| Dec | Deliver first order; get VC funds | £1.0M | Important to business; important to value |
| *Year 2* | | | |
| Mar | Sale of 25 projectors – company makes a profit for the first time | £1.5M | Important to business; very important to value |
| Jun | Sales of £15M and profits of £1M are forecast for Year 2 | £2M | Important to business; quite important to value |
| Dec | Sales and profit targets achieved; projected sales of £30M for the next year; negotiate second VC round | £10M | Important to business; very important to value |

CEC Design has changed from being a new idea to being a real business with revenues and profits, in two years. As a result, the business is worth a great deal more in Year 2 than it was at the start of its life.

## Valuation methodologies

Valuation methodologies are not as difficult as you may think, because there is no one right answer and you can legitimately use a number of different approaches to estimate the value of a business. On the other hand, it is important that you consider carefully all the issues when thinking about valuing a company to enable you to decide upon the right approach to use. In order to make an informed, if unscientific, decision, you need to understand the main approaches to valuing a business and then look at the relative strengths and weaknesses of each.

The main methods of valuing a business are:

- what someone will pay for it
- the assets of the business less the liabilities (net asset value)
- a ratio of the income of the business or the forward order-book
- price to earnings ratio (p/e ratio) – a multiple of profits based on industry sector norms
- discounted cash-flow methods looking at the future income streams
  - pay-back analysis
  - NPV
  - risk-adjusted NPV
  - internal rate of return (IRR)
- probability-adjusted methods including:
  - decision analysis
  - option-based pricing theory.

Basically, the methods of valuation are simple and uncomplicated at the top of the list, but become more and more complicated as you go down it. Option pricing at the bottom of the list, is so complex that it is strongly recommended that you leave this method to professional valuers. Let's consider these methods in more detail.

## What someone will pay

A business is only worth as much as someone is prepared to pay for it. Although this statement may seem very obvious, it is an important concept to bear in mind. What someone will pay for the business is the real and true measure of its value. If the best offer that you can get is £10 million, there is no point in valuing your business at £50 million. This value might be set, for example, by how much it would cost a rival to enter the market.

This is particularly true of new companies. However, with more mature companies that have been launched on the stock market, the value is fixed on the share price, as set by the market. The value of the company changes every day, as share prices will fluctuate. Although the value of the company is determined by the share price, the share price is influenced, in turn, by the perceived value of the company, which is in turn influenced by news about the company.

## Net asset value

This is one of the harshest methods of valuing a business. The net asset value of a business is the value of the assets that a business owns, minus the value of its liabilities, which are what the business owes to other people. For a new start-up firm, this valuation method will give a very low, or even negative value. The net asset value is a method which is commonly used for valuing businesses that are being 'wound-up' or closed down. However, in a forced sale the assets of the business may not be worth very much, and they will certainly be worth considerably less than if they were valued as part of a 'going concern'. Net asset value valuations are not appropriate for valuing new business ventures, and should be avoided. If someone suggests using this method to value your business, look for a different deal.

‘NET ASSET VALUE VALUATIONS ARE NOT APPROPRIATE FOR VALUING NEW BUSINESS VENTURES’

## Revenue-based valuation

This is a very simple, rough-and-ready method, in which the value of a business is based on a multiple of the level of sales over the current year.

This method makes assumptions about the ratio of profit to turnover, and the size of the multiple depends on the type of industry you are in and how long you have been trading for. The multiple figure will be bigger the larger the share of the market that the company controls, and can show itself able to maintain. For example, if you own a business that provides basic computer software training, an endeavour with few assets and few barriers to market entry, and have been in business for three or four years, the multiple figure would probably be about 1 or 1.5, which is quite low. Therefore, if your sales this year are likely to be £2 million, the value of the business will probably be about £2–3 million (as long as you can find someone who will be willing to pay this much for it!). An example of a company that might have a much higher valuation multiple could be a business designing and selling computer chips with a unique architecture, based on patented technology that it owns, and has proved in the marketplace. The valuation multiple of such a company could be five times or more than the previous example. ARM Holdings in Cambridge is an example of such a business, and at one point the value of this company was 40 times the sales revenue! This extremely high valuation multiple was based on the fact that the market had high expectations of a rapid growth in profits at that time. The valuation multiple has now dropped to about ten times the revenue. This was not because the level of sales or profitability changed, but because stock market sentiment moved against high-tech companies, and the expectation of future profits has become more realistic.

Revenue-based valuation is not very precise, and produces particular problems for new start-up companies, such as:

- My company doesn't have any sales yet.
- We've only just started the business and our sales are very low.
- I don't know what the 'multiple' for my industry is.
- A small change in the multiple makes a big difference in the value.
- I've just lost a big sales contract but I know things will be better next year.

The first two problems will cause the biggest difficulty for a new start-up company, because without any or little revenue, a company will not be able to value themselves satisfactorily by this method.

So, while sales or revenue multiples may be a quick and simple method for established businesses, this method has some serious shortcomings for new businesses that are yet to get off the ground.

## Price to earnings ratio

This method uses a ratio of the profits of the firm, rather than the sales figures, to estimate the price/value of the business. The price to earnings ratio is generally considered to give a better measure of the value of a company than the sales-based approach used above. As in revenue-based valuations, different industries have different multiples or ratios that are used to calculate the value. Commodity businesses will have a ratio of about 5 to 10, engineering businesses will have a ratio in the region of 8 to 12, software businesses have ratios of 8 to 15 and pharmaceuticals will have ratios of 18 to 30.

The advantages of this method are that:

- It is very quick to calculate.
- It is relatively consistent within industry sectors.
- As it is based on profits rather than sales, the resultant value gives an indication of whether the business is being run efficiently.

As a result of these advantages, this is probably the most common 'first analysis' method used by venture capital firms when they are looking at a new venture and its business plan. Bearing this in mind, make sure that you use the correct ratios for your industry, which you can check by looking at ratios for similar companies, and make sure that the expected profit figures tell the right tale.

Using p/e ratios to value a business that has been trading for some time is fairly straightforward. The weaknesses of this approach are very similar to those described above for the sales/revenue multiple valuation technique, with the principal problem being how to value a business with no sales or profits. You can see how this technique is better suited to existing 'going concern' businesses rather than new business ventures. However, that being said, this method is one of the preferred methods of valuation for VCs and you should make sure that you understand it well.

Taking the example of CEC Design Ltd, let's consider what the value of the business would have been at the end of the second year, having made a profit of £1 million on a turnover of £15 million.

A VC firm would probably calculate the value of the business at this point using a p/e of 10, which means that the value would be ten times the profits in that year: 10 x £1 million = £10 million.

For businesses with high capital needs, a measure of profit that excludes the cost of capital, such as EBITDA (Earnings before Interest, Tax, Depreciation and Amortization) is sometimes used, particularly in the US.

**Breakeven analysis** This is considered quite a useful analysis, more because of the insight that it gives you, even though it uses future projections.

Breakeven analysis estimates the time that it will take for the sales of a new business venture to pay back the costs (fixed and variable) that have been incurred. There are two points of significance: *day-to-day breakeven* occurs when the sales equal the costs. At this point, there is no profit or loss, but thereafter the company will start to make a positive cumulative cash flow. Eventually this cash flow will accumulate to the point where it can pay back the investment in the business, and there is *overall breakeven*. This time period is not only critical for the owners of a business, but also for the investors, who are waiting to get their money back and to see a profitable return. Up to the breakeven point, the business has spent more money than it has earned, which is obviously a concern for investors, even if the future

## Breakeven timing for CEC Design

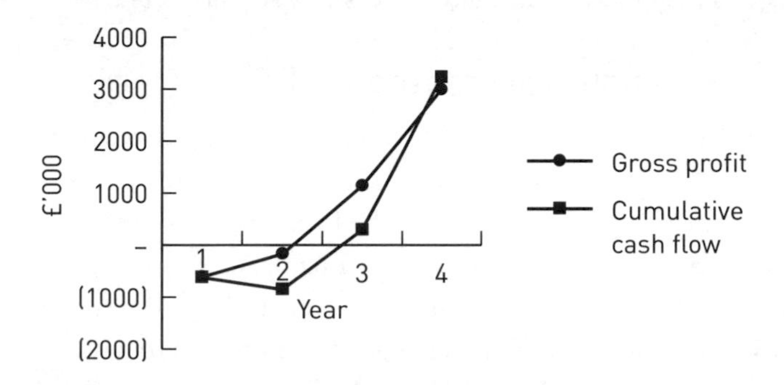


We can see from the graph that positive cash flow occurs in Year 2, but that breakeven is not until Year 3. the cumulative cash flow line is often referred to as a 'hockey stick' due to its shape – it always goes down at the beginning and then goes up steeply, crosses the 'breakeven' line and then keeps going upwards. Be aware these are only projections. Until the cash is in the bank they are works of fiction.

looks very rosy. Many investors will want to know what the breakeven timing is, and what the key factors that could affect the timing of breakeven are. When you are planning your new business venture, this is one analysis that you should make and think about the key factors affecting the timings.

In industry sectors with long development periods such as biotech or manufacturing, it can take a surprisingly and sometimes disappointingly long time for a company to reach breakeven point, and it is essential that investors are aware of this.

❛ MANY INVESTORS WILL WANT TO KNOW WHAT THE BREAKEVEN TIMING IS ❜

In order to calculate breakeven timing (and most discounted cash-flow techniques) you need to work out the cash flow for the business, which includes both cash out for expenditure and investment, and cash in from sales and profits, over each relevant period of time. The time period that you use will depend on the business, and how fast it is changing or growing. For example, with a new business, the cash flow forecast should be calculated at least every quarter and probably every month and in a really fast changing new business you may need to calculate the cash flow every week.

## Discounted cash-flow methods (DCF)

All the methods discussed so far treat money as though it is worth the same now as it will be in the future. As discussed in Chapter 5, this is not the case, as money that you might earn in the future has less present value than the money that you have in your pocket or bank account today. Discounted cash-flow methods look at the cash flows of your business, and adjust the value of future cash flows. This concept can be applied to a number of different valuation approaches, each of which tells you something slightly different about the business. These are:

- discounted cash flow
- net present value analysis of profitability (NPV)
- risk-adjusted NPV
- internal rate of return (IRR).

Before we get into the uses, advantages and disadvantages of these methods, let's go quickly through the maths of discounted cash flow, which although they may sound intimidating, are really quite easy.

‘FIRST, YOU NEED TO AGREE A ‘DISCOUNT RATE’, WHICH IS THE RATE BY WHICH MONEY IS REDUCED OVER A SPECIFIC PERIOD OF TIME’

First, you need to agree a ‘discount rate’, which is the rate by which money is reduced over a specific period of time. The discount rate is very similar to the interest rate that you have to pay on any money borrowed from a bank or on your mortgage. A discount rate of 10 per cent would mean that money received next year is worth ten per cent less than money in your pocket today. For example, £100 received in one year's time would only be worth just over £90 in today's money.

When valuing a business, you will need to take into consideration the expected cash flow and profits over a period of time, maybe three, five or ten years, although the further into the future, the more uncertain these numbers become. However, it is important to be able to understand this future cash in present value for these amounts to have any meaning. The further into the future that you predict, the less it will be worth in today's money. As we saw in Chapter 5, the formula to calculate this is:

$$DCF = \Sigma I/(1+r)^n$$

Where I is the cash flow in some year, r is the discount rate for the period, n is the number of years in the future.

Following this rule, cash earned in the present year is worth its face value, cash earned next year is worth a little less (depending on the discount factor), and profits earned during the year after are worth even less than that, as they are earned even further into the future.

## Discount factors

| *Rate* | *4%* | *8%* | *10%* | *15%* |
|---|---|---|---|---|
| Year 1 | 0.962 | 0.926 | 0.909 | 0.870 |
| Year 2 | 0.925 | 0.857 | 0.826 | 0.756 |
| Year 3 | 0.889 | 0.794 | 0.751 | 0.658 |
| Year 4 | 0.885 | 0.735 | 0.683 | 0.572 |
| Year 5 | 0.822 | 0.681 | 0.621 | 0.497 |

This table allows you to short-cut the present value calculation, by looking up the discount factor for a given rate and period.

Thus the present value (PV) of £100 in four years' time at a discount rate of ten per cent is £100 × 0.683 = £68.31.

Looking at the revenue figures for CEC Design Ltd we have

| *Year* | *1* | *2* | *3* | *4* |
|---|---|---|---|---|
| £'000 | 300 | 2,100 | 5,700 | 14,700 |

This corresponds to a DCF of about £6 million. At the end of the first year a VC might discount this by a factor of 4, and using an industry ratio of 1:1 from revenue to profit, estimate a valuation of around £1.5 million.

There are some important characteristics of discounted cash-flow techniques. First choosing the right discount rate is very important, as if you use a discount rate that is too high, the value of future profits will be radically reduced. Second, don't be fooled into the belief that this is a magic formula; just because you have used mathematics to estimate the value of your business does not mean that future profits are guaranteed. Future profits rely on the performance of you and your team and a host of random factors such as chance, market forces, political climate, etc. Anything more than a couple of years into the future is likely to be fiction.

‘ANYTHING MORE THAN A COUPLE OF YEARS INTO THE FUTURE IS LIKELY TO BE FICTION’

The advantages and disadvantages of discounted cash-flow methods are as follows.

*Advantages*:

1 It can give a more accurate estimate of the value of the business as more data is considered.

2 It indicates to financiers and venture capitalists that you know something about money and valuing businesses.

3 It places more emphasis on future rather than current income, which is better for early-stage businesses.

*Disadvantages*:

1. The 'change in future value' calculated by the discounted cash-flow methods will almost certainly be smaller than the 'margins of error' in your future profits calculations, which means that the overall result will be inaccurate however accurately you calculate the discounted cash flow.
2. Venture capitalists are very aware of such inaccuracies, and as a result discount the result heavily.
3. Time can be wasted calculating and worrying about discounted cash-flow techniques which would be better spent working on your business strategy and marketing plan.

Looking at the profit figures for CEC Design Ltd we have:

| *Year* | *1* | *2* | *3* | *4* |
|---|---|---|---|---|
| Gross profit | (656) | (199) | 1,103 | 2,996 |

This corresponds to an NPV of about £2 million.

At the end of the first year a VC might discount this by a factor of four, and using an industry ratio 1:10 of price to earnings (p/e ratio), estimate a valuation of around £5 million

This figure is high because our company is predicted to be rather more efficient than average, with a gross profitability of around 30 per cent, as most of its income is predicted to come from licensing its technology. If it had a more normal profitability of about ten per cent, its estimated value would be closer to £1.6 million, and a prudent VC might make this correction

**Net present value (NPV) of profitability** Net present value techniques are a way of dealing with the time value of money. NPV of profitability looks at the positive and negative discounted profits for a project over a specific time period. By adding these values together, a 'net value' figure is produced for the period, which is at present-day value. The result is a single number representing, in present-day value, the profitability of the project as a whole.

NPV of profitability analysis is a very useful technique, which is often used by large companies and banks for valuing projects and future investments. This is because it allows you to compare projects with different time periods, investment requirements and income streams, and thereby decide which offers the business the best return. When evaluating investments venture capitalists and bankers sometimes use NPV of profitability analysis, although such firms will more often look for simpler valuation methods. Because profit is the difference of two large numbers – the revenue and the costs – it is numerically unstable, and many things can go wrong with the prediction. However, as a technique for valuing a new business, NPV of profitability is less useful to value a single company in isolation, and the method's strengths lie in its ability to compare projects. In effect, this method is very similar to that of valuation by taking a multiple of profits, but with a bit of mathematics added in which make it feel as if it is more than just a 'good guess'.

The advantages and disadvantages of NPV of profitability are as follows.

*Advantages:*

1. It gives a value for the business over a given time period.
2. It is an approved financial approach to valuation.
3. By your use of NPV, bankers and venture capitalists will see that you understand valuation methods.
4. It allows you to compare the value of two or more competing business ideas (if you are lucky enough to have them).

*Disadvantages*:

1 It can foster over-confidence in the achievability of the valuation.

2 The calculation of NPV can distract the business team from more important tasks such as business strategy and marketing plans.

3 The selection of discount rates and time periods will have a major effect on the calculated value.

NPV is an important method to understand and is probably the most useful of the more 'sophisticated' valuation methods. Because it provides a good balance between being too simple and too complex, it is recommended that you spend a little time getting to understand how to value your business using this approach.

❝NPV OF PROFITABILITY ANALYSIS IS A VERY USEFUL TECHNIQUE, WHICH IS OFTEN USED BY LARGE COMPANIES AND BANKS FOR VALUING PROJECTS AND FUTURE INVESTMENTS❞

## Probability-adjusted methods

With all of the methods outlined above, you also need to be aware of the risk that things may not turn out as expected, and that the real value of a business may be less than that predicted, as these risks are not actually taken into account.

However, there are methods for valuing businesses that estimate some of the risks, such as the risk that the market may not be as big, or may be bigger, and that the competition may be stronger or weaker than assumed. These are called probability-adjusted methods and, while they may be more complicated than other methods, they do address the issues of risk. However, whether the additional information about the business is worth the complicated calculations is not clear, as much of the calculation is in fact based on guess work. The result is only as good as the estimates that go into it, and no amount of mathematical fiddling can help. Garbage in, nonsense out.

❝GARBAGE IN, NONSENSE OUT❞

**Decision analysis** This method assumes that there are decisions or options that will occur that will affect the value of your business, and by assigning probabilities to these alternative outcomes, different valuations may be determined. One way to demonstrate this method is with a matrix that looks at the probability of success of a new technology (competitive success), and the probability of the size of the market.

## Decision analysis matrix for CEC Design Ltd

| | **CEC lags behind competition: Market share 5% with probability 20%** | **CEC ahead of competition: Market share 20% with probability 70%** | **No competition: Market share 100% with probability 10%** |
|---|---|---|---|
| **Market expands faster than expected:** | Sales of £25m<br>Probability 2% | Sales of £100M<br>Probability 7% | Sales of £500M<br>Probability 1% |
| **Market size: £500M with probability 10%** | Poor product in big market | Good product in a big market | Only product in a big market |
| **Market grows as expected:** | Sales of £5M<br>Probability 14% | Sales of £20M<br>Probability 49% | Sales of £100M<br>Probability 7% |
| **Market size: £100M with probability 70%** | Poor product in expected market | Good product in expected market | Only product in expected market |
| **Market fails to develop quickly:** | Sales of £500K<br>Probability 4% | Sales of £2M<br>Probability 14% | Sales of £10M<br>Probability 2% |
| **Market size: £10M with probability 20%** | Poor product in weak market | Good product in weak market | Only product in weak market |

Values can be calculated for each of the options (and these could be DCF-based values if you want to spend a lot of time on the arithmetic) and, depending on your views as to which of these outcomes is most likely to take place (another requirement for judgement-based guess work), you could estimate the most likely worth of a new business venture. However, as you can see, this will involve a considerable number of estimates to be made (guesses to you and me) and then a lot of arithmetic to make the analysis.

You will probably have guessed that I don't think the extra pain of analysis is worthwhile in most cases. Effectively our VC was doing a short-cut version of this analysis and compensating for future uncertainty by applying a discount factor of four to the DCF results in the example. Time and experience have taught them that this is an appropriate scaling factor for this sort of business. However, it may well be that in certain cases such a review of options will cause the business team to think about the business in a new way, or to realize that there are some important factors that they have not taken into account. If such review results from an analysis, then it will probably have been well worthwhile, and may have saved time, money and effort from being expended in the wrong direction.

‘YOU WILL PROBABLY HAVE GUESSED THAT I DON'T THINK THE EXTRA PAIN OF ANALYSIS IS WORTH-WHILE IN MOST CASES’

**Option-based pricing theory** Option-based pricing is a further development of probability-based valuation, and provides the most sophisticated valuation approach. One version of option-based pricing is the prize-winning Black-Scholes model mentioned earlier. Option-based pricing estimates a range of probabilities involved at each step during the development of a project or business and, using statistical methods, such as a 'Monte Carlo simulation', estimates the value for each alternative option. You can even make guesses as to the accuracy of your guess, and figure that into the equation. This method is used by some stockbrokers to value option dealing, and by some pharmaceutical firms to compare the investments in major drugs projects. However, it is not really an appropriate method to value smaller new businesses, as it is very complex, and requires a lot of mathematics, computing power and time to work through the analysis. At the end of the day it is still only as accurate as the assumptions that you made during the process.

## Summary

Valuation is very important for you and your business and you need to address this early in the process of starting your new venture, even though it won't be worth very much at an early stage. There are many methods that you can use to value your business, but remember these rules:

- Your business is only worth what someone will pay for it.
- Buyers and sellers often have different ideas as to value.
- Simple methods for valuation are probably more cost-effective of your time than more complex analyses.
- The value of your business will change with time – hopefully going upwards.
- The timing of bringing in new investors can be critical.
- Make sure you get any valuation agreement or 'deal' in writing – verbal contracts aren't worth the paper they are written on.

## Exercise

**one** Find the accounts for a publicly quoted company, such as ARM Ltd. Do your own valuation, not looking at the share price, and then compare your result with the value of the company given by the share price. How do you explain the difference?

## Further reading

Copeland, T., Koller, T. and Murrin, J. (1994) *Valuation: measuring and managing the value of companies*. 2nd edn. Wiley & Sons

'Valuing your company' **www.startups.com**

**www.bvca.co.uk** British Venture Capital Association

**www.entreworld.org** Entrepreneurs Resource Centre

**www.solutionmatrix.com/**
(Business case analysis tools from a US consultancy)

# 18 Exit routes

'If I am not for myself, then who will be for me?'

(Rabbi Hillel, Mishnah, 14–15)

IN THIS CHAPTER WE DISCUSS HOW TO TURN WHAT you have built into cash for yourself and your investors, and draw some conclusions. The rest is then up to you.

## The wild-flower model

Start-up companies are a bit like wild flowers in a weed patch. Bright sparks of entrepreneurs are the seeds. With luck, the start-up companies, or wild-flower plants, will find a fertile niche of a ready market where they can grow, encouraged by the sun of venture funding. As with a plant, one of several things can happen:

- They grow and colonize neighbouring sites.
- They find a small niche and manage to survive, but do not thrive.
- They are crowded out by the competition and die.
- Some large entity comes along and harvests them.
- They come to the end of their natural life and fall over. The bright sparks of their ripe seeds, employees who have been attracted to work for them, start new plants, and the cycle repeats.

The model of successful companies seeding new enterprises is particularly apparent in places like Silicon Valley in California, where, for instance, most of the semiconductor companies can trace a linear heritage to a company called Fairchild ('The Fairchildren'). Similarly, in Cambridge UK, there is a cluster of computer-aided design (CAD) companies that can trace direct descent from the decision of Professor Wilkes at the University's Computer Laboratory to buy an early graphics display, and make it freely available to researchers in the laboratory.

Exiting is part of a natural process, but there needs to be business logic behind it for all parties concerned. Getting rich quick doesn't qualify.

If your primary aim is to sell or float your company early for lots of money, then you will almost certainly fail. The company has to be a valid business in its own right, and you need to be committed to it, and turned on by it.

'IF YOUR PRIMARY AIM IS TO SELL OR FLOAT YOUR COMPANY EARLY FOR LOTS OF MONEY, THEN YOU WILL ALMOST CERTAINLY FAIL'

It may be that you are quite content with the way your company is. It supports your lifestyle, and you are happy to potter along, maybe passing it to your son or daughter when you eventually retire. If so, you are a lucky person. Most companies have their own dynamic and lifetime, and sooner or later the time comes when change is needed, or the people involved want to go and do other things.

If you are considering exiting a company, you need to consider more than just the financial gain. In any exit, it is important to consider the effect that it will have on the people involved, such as the employees, their families, and even the community in which the company operates. This can make for some tough decisions, as after all, these were the people who made the company successful. Sometimes an exit can be good news, and the floatation of the company or buy-out enriches the people who helped to make it a success. For example, the success of ARM Ltd has enriched of most of its early staff, including some of the secretaries, receptionists and cleaners who stayed with the company. However, in other cases, the exit of a company is not good news, and the closure of a plant can decimate a community. Whatever the final outcome, it is important that you consult all concerned as early as possible, to brief and prepare them for what is likely to happen. There may be local and governmental agencies that can assist with this process. If the enterprise is of any size, or at all newsworthy, you had better get your press and PR position sorted out early too, as the press will be after you regardless of whether the news is good or bad.

## Acquisition

Acquisition or trade sale is a common fate of a start-up, especially in industries that are going through a consolidation process. In many ways, acquisition is the cleanest, quickest and most satisfactory exit route. Indeed, in the case of software companies, some people reckon that acquisition by Microsoft is the only alternative fate to death!

In an acquisition, the purchaser can be:

- a larger group acquiring expertise
- a rival, possibly from overseas, acquiring market share
- a supplier or distributor, who are vertically integrating
- a financial house stripping asset value, buying the company to break it up, where the parts or the assets, such as freehold property are worth more than the whole as a going concern.

Acquisition can be very positive, by, for example, providing more resources, or bigger market opportunities for the enterprise. Even asset stripping and break-up can inject some reality, and allow parts of the business to flourish on their own without the albatross of the organization around their necks. Of course, there will be culture clashes, and there may be a certain amount of rationalization. However, if this is handled carefully and sensitively, the pain need not last long, as the benefits will soon become obvious.

Acquisitions don't happen by chance. Like a marriage, people must somehow meet and have a period of courtship before tying the knot. Often, some sort of business relationship is formed first, such as through a joint venture, working together on a project as part of a consortium, being a supplier or distributor, or some form of minority investment. If eventual acquisition is your aim, it might be wise to start flirting with possible partners, and to remove any barriers, such as ensuring that your systems are compatible with the potential partner's, or that your software runs on Windows operating systems, for example. Note that this process can take many years, before you are in a position to sign a deal.

‘ACQUISITIONS DON'T HAPPEN BY CHANCE’

The exception to this is a distress sale, where the company is actively marketed as an alternative to going out of business. In such a sale, time is not on your side.

Any acquisition will need the services of professionals to negotiate terms, make sure your books are straight and to make sure that all other necessary procedures are done with due diligence.

## Flotation

Flotation is the process of selling a proportion of the shares of the company to the general public (primary market) at a fixed price. The shares are then subsequently traded on a recognized market (the secondary market) such as the stock exchange at a floating price.

Flotation is mostly a way of raising money to fund the company's expansion. However, as it places a market value on shares in the company, it can provide an exit route for investors and founders. There are, however, usually lock-in periods for the founders and senior staff, typically of one year, but can be more or less, and there may be other limitations to ensure an orderly market.

Flotation is expensive, and may involve professional fees of many millions, typically ten per cent of the money raised. It is not worth going to the market unless you are sure that the company is capable of raising significant capital. Recent years have seen the development of junior markets with marginally less stringent entry conditions, such as AIM in the UK, NASDAQ in the US, Le Deuxième Marché in France and Die Neue Mark in Germany. The original intention of these markets was that they would offer a higher return and a higher risk profile, making it easier for high-growth companies to come to market earlier in their life cycle. In practice, there is little advantage in cost to floating on a junior market and, with the exception of NASDAQ, the junior markets have not been as successful as was hoped. Other higher-risk markets exist, such as EASDAQ and the Vancouver stock market, but these are mainly for specialist areas, for example, mining and mineral exploration stocks on the Vancouver market.

A company needs a trading record of at least three years before flotation on most markets. It may also take a year or more to prepare for flotation once the company has decided to take that course. Flotation is risky, since the market may not buy the stock. This is where the underwriter, who will guarantee to buy at a particular price, comes in handy. However, even with this insurance measure, if the market moves against the stock and its value, the value of the company and the investor's investment may decline severely in the secondary market. This has been the fate of many dot coms. During a recession, the market prefers companies with a solid track record of profitability today, rather than a promise of jam tomorrow.

‘FLOTATION IS RISKY, SINCE THE MARKET MAY NOT BUY THE STOCK’

Flotation has other disadvantages. In effect, it will mean that the company will gain another product group, the shares, in addition to the actual products of the enterprise. The stock will need as much attention as any other product, requiring briefings to analysts and a continuous supply of good news to the market to maintain the share price. Maintaining good relations with the City takes up a lot of management time. Note also that financial, reporting, audit and other requirements are much stricter and more onerous for publicly listed companies than for private companies.

## Buying and selling shares

This is based mostly on ~~UK jurisdiction~~, but similar rules govern share trading in most countries.

**Advertising shares** Although it is legal to buy, hold and sell shares privately, it is illegal to advertise them for sale unless you are a practitioner, such as a broker, and are suitably licensed by a quasi-governmental body, for example, the Financial Services Authority in the UK, or the Securities and Exchange Commission in the US. This law is designed to prevent the sale of shares in fraudulent companies to the unsuspecting.

This law means that you cannot put up a website offering shares in your new and exciting venture, or even legally send e-mails to all your friends asking for contributions. This leads to something of a chicken-and-egg problem, and there is an exception in the law to allow for the distribution of a prospectus to a small number of people (usually no more than 20 or so) who have somehow managed to formally ask to be informed. Of course, asking them to ask you counts as advertising, and they therefore need to somehow guess that you might be willing to sell shares in your company.

## Options and futures

Options and futures are contracts to buy or sell something for a fixed price at a designed future date. A futures contract must be completed as specified, and an options contract allows the purchaser the option of completion or abandonment. Options cost more than futures. Options and futures contracts may be traded in their own right.

Futures and options contracts were originally developed in the commodities markets. For example, a manufacturer might wish to ensure a future supply of, say, copper, and therefore might purchase their requirements ahead of time. They are also used in the currency markets to hedge against currency fluctuations. Options are often granted to employees as a tax-efficient way to benefit from an increase in the company's value.

The cost of an option is typically ten per cent of the strike price, the cost of the underlying security. It can thus be used to gear or hedge an investment. For example, if you are convinced that the market will rise, you can purchase a call option on the index. Call options allow you to purchase at a certain price at a fixed time in the future. If the market has risen (a *bull* market), you can exercise the option, buy the shares at the fixed lower price, and then sell them at the higher market price, making a tidy profit. You will get back ten times more than what you would have received if you had just invested in the underlying securities. However, if the market drops (a *bear* market),

**Markets** The *primary market* is where the shares are first sold to the public, normally through an *Initial Public Offering* (IPO). The conduct of an IPO is governed by many rules and regulations, and you will need skilled and expensive professional help to guide you through this process. Before the shares are traded, an initial *strike price* is established, which is the price at which the company will sell the shares.

you lose all your money and get nothing, whereas you may still have had some value if you had invested in the underlying security.

Future, and options allow the trader to make money in a falling or bear market. If you are convinced that the market will drop, you can take out a put option, which is an option to sell at a fixed price. If the market drops below that price, you can buy shares and sell them at the higher fixed price. Of, course if the market goes up, you lose the purchase price of the option, and do not have the comfort of owning the underlying security.

Put and call options for different values and dates can be combined in various ways to give optimal yields for differing scenarios and levels of risk. Valuation of futures and options contracts can involve sophisticated financial and statistical modelling.

Trading in futures and options, because of the gearing, allows you to make and lose money very quickly. You can lose many times more than your original investment. It is a form of high-stakes gambling, and best left to the professionals. However, even professionals can get it wrong, as the Leeson affair that broke Barings Bank demonstrates.

Options can be used to hedge the value of shares, for example during the lock-in period after you have sold the company, but before you can sell the shares. Ask your financial adviser to explain about cashless collars for this use.

## Other financial instruments

### *Spread betting*

Just as you can bet on the horses, you can bet on the markets. You can even gamble on the markets over the internet. Various firms offer spread betting,where, for example, they quote a target range for a stock exchange index such as the FTSE 100 for some future date, such as tomorrow's opening. If you think the predicted target is too low, you can buy an 'up bet', or if you think it is too high, a 'down bet'. Bets are calculated on a certain amount per point, such as £10/point. If the market opens at a level in accordance with your bet on the specified date, you will win the difference multiplied by the amount you bet per point. The same applies if you are wrong, and you will have to pay the difference multiplied by the amount per point. As this activity is a bet, not an investment, it is tax-free in the UK and US. Unless you specify a *limit bet* your losses can be unlimited, and you can lose lots of money fast. Bookmakers are not known to be philanthropists. Unless you are very sure that you know what you are doing, steer clear of this game and be very wary of the potentially unlimited downside. For any investment, it is extremely foolish to risk more than you can easily afford to lose.

### *Margin trading*

Another form of gearing is margin trading. Your broker may, for a fee, lend you money against the security of the stocks you have purchased and lodged with them. For example, if you have purchased £2,000-worth of stocks, your broker may be prepared to lend you 50 per cent, or £1,000. Thus for your £2,000 investment, you can buy £3,000-worth of stock. In rising or bull market, all well and good. Suppose the stock rises by 50 per cent, you will get £4,500, which is more than doubling your profit. However, in a dropping or bear market your money disappears even faster. If the market drops 50 per cent, your investment is now only worth £1,500, of which the broker will still want his money back, leaving you with £500, which is more than doubling your loss. Worse, the broker will only be prepared to lend you 50 per cent of your investment, or £250, so will make a margin call on you for £750, which you must pay, or sell your investment to repay them, even if the time is not good.

A *secondary market* is established when the shares are admitted to be traded on a public exchange, along with the IPO. The secondary market consists of further trading by people who bought shares in the primary market, and are now selling to others. If the exchange used is the principle stock exchange for the country, the shares may be admitted to the country's *official list*.

‘A SECONDARY MARKET IS ESTABLISHED WHEN THE SHARES ARE ADMITTED TO BE TRADED ON A PUBLIC EXCHANGE, ALONG WITH THE IPO’

Shares can now be traded. Again, most jurisdictions do not allow shares to be advertised, except by authorized personnel, and therefore most dealing is done via a broker or equivalent. Recent times have seen the rise of internet brokerages that provide easy access to the market.

'Market makers' are brokers who take a position in the stock, and are committed to providing a continuous price at which they will both buy and sell the stock, at least at the normal market volumes. However, this service is in return for certain privileges, such as having the opportunity to bid or offer on all trades. Market makers provide liquidity and inertia, and act as the flywheel for the market. The difference between the buy price and the sell price is known as the spread, and can be quite large for illiquid and volatile stocks. A good broker, negotiating directly with the market maker, can often trade inside the spread. Indeed most stocks, outside the top 300 companies or so, are illiquid, and for very large trades (say ten times the normal market volume), all stocks are illiquid.

Effectively, this results in a two-tier system, with the normal market mechanisms applying for small and everyday trades, but a specialist market emerging, sometimes known as the upstairs market, where large trade deals occur by direct negotiation. Of course, normal exchange rules mean that the trade must subsequently be reported to the market, but the price discovery and negotiation mechanisms are quite separate.

‘IT IS ILLEGAL TO USE OR DIVULGE INSIDER INFORMATION ABOUT A COMPANY THAT HAS NOT YET BEEN ANNOUNCED TO THE MARKET’

**Inside knowledge** One of the tenets of a free and fair market is that all participants have access to the same information. It is illegal to use or divulge insider information about a company that has not yet been announced to the market. This means, for example, that there are ‘closed periods’ of trading, such as before publication of results, where the directors and senior staff of the company, and their families and associates, are prohibited from trading in the shares of the company. The exchange can, and does, scrutinize any trade within these closed periods.

If you have got to this stage it means you have potentially cashed out, and won. You will be surrounded by professional advisers, and don’t need this book any more.

## MBO

A management buy-out is where the senior staff purchase the company from the investors. Since the staff have been the ones actually running the company, there is no learning curve to go through. Financial assessments are also relatively easy, since the company’s track record will be based on the same team. As a result, it is usually quite straightforward to raise new investment from banks and venture funds to fund the purchase, with the staff only contributing a nominal amount in actual cash. Often, a few years down the line, MBO companies float in turn, or are acquired.

There are many variants of the MBO theme, such as, for example, a management buy-in, where an experienced management team purchases another company, replacing the management.

## Liquidation

The directors of the company may just tire of the business and decide to wind it up, selling the assets and, after paying the debts, returning any money left to the shareholders.

Clearly this route would not normally raise as much money as selling the whole business as a going concern, but if the business is not profitable, and has little prospect of being so, it may be the only option. In exceptional circumstances, the value of the assets may exceed the value of the business. This could apply to situations, where, for example, the assets are undervalued, or where staying in business may involve some contingent liability.

In situations where the sale is forced, for example in bankruptcy, assets may fetch only a small fraction of their book worth. For instance, the market for second-hand computers, even if still usable, is small, and the value of an incomplete or specialist piece of software is virtually zero, even if it cost many hundreds of thousands of pounds to build.

## Bankruptcy

If the company cannot pay its creditors, and is unlikely to be able to do so in a timely fashion, then it must be declared bankrupt. Failure to do so is a criminal offence, and the directors can go to prison. A director's liability can persist for up to a year after they have resigned. If you are a director of a company and suspect that your company is, or is likely to be, in this position, seek professional help from your accountant and legal adviser as soon as possible. There are things that they can do to help.

Note that there are various get-out clauses. Although the company may be technically insolvent, the directors may have reasonable belief that they can raise the money in time, or that a debtor will pay in time and that they can trade out of the position. The directors may be able to come to some arrangement with their major creditors, who might agree to wait for their money. In these cases, the company can continue to trade, but any Board meeting should make careful note in the minutes of the circumstances. 'The directors, having carefully considered all the circumstances, and after discussion with the company's bankers, re-affirm their belief in the company's ability to raise sufficient money to met the company's obligations in a timely fashion' might be one version of such a note.

A company can be declared bankrupt voluntarily by its directors, or involuntarily by its creditors applying to the courts. Failing to pay or make provision for taxes, such as payroll or VAT, is one common cause of involuntary bankruptcy, and the taxman is unforgiving and relentless.

‘WHEN A COMPANY IS DECLARED BANKRUPT, THE INTERESTS OF ITS CREDITORS TAKE PRIORITY OVER THOSE OF ITS SHAREHOLDERS’

When a company is declared bankrupt, the interests of its creditors take priority over those of its shareholders. Normally, a receiver is appointed to salvage as much as they can, and will either make a trade sale or, in extreme situations, liquidate and wind up the company, paying the creditors and returning any surplus to the shareholders. However, there is rarely any surplus after the receiver's and liquidator's fees.

Most jurisdictions provide some form of half-way house, where protection is given from the company's creditors, to allow the company to continue as a going concern. This will provide the company with an opportunity to trade out of trouble, raise more cash or find a buyer. In the UK, this protection is part of an arrangement known as administrative receivership, and in the US, this provision is included under Chapter 11 of the bankruptcy regulations.

Clearly it is better not to 'change your life and meet strange new people' by going bankrupt if it can be avoided. Get help as early as possible. There are lots of people who can help, starting with your bank manager (sorry, relationship manager!). The bank would much rather that you get out of trouble than have to foreclose your company. Your bank may be willing to provide assistance in the form of soft loans or repayment holidays, and may be able to put you in touch with business reorganization specialists. Other people who can also help are your accountancy advisers, and even your major creditors. For instance, explain to your creditors that if they are patient and can wait a bit longer for their money, they will stand a better chance of getting it, as if you are pushed over the edge, they might end up getting nothing. Likewise, your customers may be able to speed up their payments, or pay you in advance or early, rather than lose you, along with any deposit that they may have already paid for partly completed work. When the going gets tough, the tough get going. There is lots to do. It's a time for action, not sitting mesmerized until disaster strikes. There is a storm coming, and you had better prepare.

Some people are unable to admit such looming disaster, not to themselves, their bank or their work colleagues. These are the people who most need help and counselling. If need be, send them home on gardening leave, while sorting out the mess they have left behind.

Company failure is no longer the social disaster it once was, providing that it was not due to dishonesty. Sometimes the cards are just stacked against you, and nobody could have succeeded with the hand that you were dealt. Indeed, some venture backers prefer to see one or two failures on a CV, as it shows that you have tried and, of course, you can't learn from your mistakes unless you have made them.

## Social responsibility

The people who start new enterprises are building the future for all of us. It would be remiss to finish this book without writing a few words on social responsibility.

> "BEING FOR MYSELF, WHO AM I?"
>
> (RABBI HILLEL, MISHNA, 14–15)

Your enterprise will affect many more people than just yourself, and your fellow directors. It will change the lives of your employees, their families, your family, your community, and potentially the lives of your suppliers and distributors as well. It is worth pausing occasionally to check that these changes will be for the better.

The things that you invent and the products that you bring to market will affect people's lives too. Will your product make the world a better place?

Do you want to live in the world with that product in it? I was involved with the design of the BBC microcomputer, an early home and educational computer, and as a result of the development we inadvertently blighted the UK with the widespread teaching of BASIC as a first computer language. With hindsight, BASIC can be accused of breaking most of the rules of good programming language design, and teaching the wrong thought patterns. I'm sure you can make your own list of things that the world would be better without, ranging from instruments of torture to irritating mobile phone ring tones. Will your new product be one of them?

> 'WILL YOUR PRODUCT MAKE THE WORLD A BETTER PLACE?'

## Spending the money

Now you have made all that money, what are you going to do with it? There is only so much money that you can reasonably spend – after all, you can only wear one set of clothes at a time, and there is only so much that you can eat and drink. You will soon get tired of sitting around all day.

Many successful entrepreneurs become serial entrepreneurs, investing their time and money in new enterprises. Others act as angels, investing in new fledgling businesses, sharing the risk, but letting others earn money for them. There are always charitable and good works to do, and there are many enterprises such as the Prince's Business Trust, where your expertise could help others up the ladder. Your alma mater or your local university will almost certainly welcome help, and may have an industrial fund, or even an incubator for new businesses that needs help and experience.

## Conclusion

> 'IF NOT NOW, WHEN?'
>
> (RABBI HILLEL, MISHNAH, 14–15)

Starting a new enterprise is an act of faith. It has the potential to generate wealth for the few, and employment for the many. It can be vastly enjoyable, and immensely frustrating. It is likely to enhance rather than damage your CV. If you don't do it, you will be left always wondering whether you could.

Now it's your turn. Are you going to just sit there?

## Exercise

**one** Take a pen and nice new notebook, or start a new file in your wordprocessor, and turn back to Chapter 4, 'Writing the business plan' ...

## Further reading

The Cambridge Phenomenon Revisited (2000) Available from Segal Quince Wickstead and Partners (**www.sqw.co.uk**)

Freiberger, P. and Swaine, M. (1999) *Fire in the Valley*. McGraw-Hill (The making of the Apple Macintosh.)

Gates, B. (2000) *Business @ the Speed of Thought*. Penguin Books

Klein, A.D. (1998) *Wallstreet.com*. Henry Holt and Company (By the founder of Wit Capital.)

Komisar, R. (2000) *The Monk and the Riddle*. Harvard Business School Press (Business start-up in Silicon Valley.)

**www.londonstockexchange.com/** London Stock Exchange

**www.nasdaq.com** NASDAQ

**www.easdaq.com** EASDAQ

# Index